WITH THE "AURORA" IN THE ANTARCTIC, 1911-1914

WITH THE "AURORA" IN THE ANTARCTIC
1911—1914

BY
JOHN KING DAVIS
Extra Master
Lieut. Commander R.A.N.R., F.R.G.S.
Chief Officer "NIMROD," British Antarctic Expedition, 1907-9
Master "AURORA," Australasian Antarctic Expedition, 1911-14
Commander "ROSS SEA" Relief Expedition, 1916

LONDON: ANDREW MELROSE, LTD.
3 YORK STREET, COVENT GARDEN, W.C.2

To
PROFESSOR T. W. EDGEWORTH DAVID, C M G , D S.O , F.R.S
IN ADMIRATION OF HIS SERVICES
TO ANTARCTIC EXPLORATION

I am indebted to the Council of the Royal Geographical Society for permission to reproduce maps which have already appeared in the "Geographical Journal", to Mr. William Heinemann for permission to reproduce matter which has already appeared in the "HOME OF THE BLIZZARD", to my shipmates whose photographs illustrate this volume

J. K. D.

PREFACE

I VENTURE to hope that this record of the voyages of the *Aurora* in Southern Waters, written from a sailor's point of view, may be useful to future explorers of the Antarctic Coast Line, and not without interest to the general public.

To Sir Douglas Mawson, the Organiser and the Leader of the Expedition (1911-14) I desire to tender my warmest thanks for his unvarying support, at all times.

I have difficulty in expressing how much the advice and encouragement of Professor David, of Sydney, and Professor Orme Masson, of Melbourne, meant to me in carrying on the work of the *Aurora* during Dr. Mawson's absence in the Antarctic, whilst I wish to thank Dr. H. R. Mill and Dr. W. S. Bruce for their advice and support in this country.

I am also indebted to Captain A. Mclean, M.C, A A.M.C, the senior medical officer at the Main Base, Adelie Land, for his assistance in reading the proofs of this book.

Finally, whatever measure of success was achieved in the special work assigned to the *Aurora* is due to the loyal co-operation of officers and men, who *did their best* in the interests of the Expedition.

JOHN KING DAVIS

AUSTRALIA HOUSE,
  LONDON,
    August, 1919

INTRODUCTORY PREFACE

THOSE interested in Antarctic Exploration will be glad of the opportunity, offered by this volume, of following the voyages of the " S.Y Aurora " as detailed by Lieutenant-Commander John King Davis, who so ably piloted the good ship on many voyages through thousands of miles of ice-strewn seas. As the author is less given to the writing of his exploits than to their performance, it is the more gratifying to receive this account from his pen

The story deals, for the most part, with those cruises when the " Aurora " was manned as the exploring vessel of the Australasian Antarctic Expedition, 1911–14. During that time the ship's company performed wonders with the good old vessel, an appreciation of which will be gained by a perusal of Davis's modest story. In the popular official account of the adventures of that Expedition, written under the title of *The Home of the Blizzard,* in order to embrace all the various activities of the enterprise, so much ground had to be covered that the voyages of the ship were but briefly touched upon. The present volume, therefore, fills a gap which would not otherwise be made good without reference to the extended Scientific Publications which, on account of their ponderosity, are not likely to appear on the shelves of popular libraries.

The present opportunity cannot be allowed to pass with-

INTRODUCTORY PREFACE

out referring to Lieutenant-Commander John King Davis himself—the man. Readers may remember that both Davis and I were comrades together as members of the Shackleton Antarctic Expedition of 1907-09. Already at that time, at the early age of twenty-four, we found him possessed of an extra-master's certificate and with years of hard nautical experience behind him.

Serving as Chief Officer during the main Antarctic voyages of the " S.Y. Nimrod," his ardent enthusiasm, capacity for hard work and earnest loyalty went far towards the success of the " Nimrod's " operations. For the return voyage, from Sydney to London, Sir Ernest Shackleton appointed Davis in command of the vessel. In which capacity, notwithstanding that it was then midwinter, he successfully voyaged far within the iceberg zone of the Southern Pacific Ocean, almost to the Antarctic Circle, in order to verify or disprove the existence of certain doubtful islands which up to that time appeared on the charts of those regions.

Davis emerged from the Shackleton Expedition as a navigator of known valour and ability, though in years still quite young. Like his namesake, the great Arctic pioneer of Elizabethan days, he had earned the reputation that beneath a rugged exterior was a God-fearing man, kind, trustworthy and courteous.

When I set myself the task of organising an Australasian Antarctic Expedition, I counted upon the co-operation of John King Davis as Second-in-Command, to take charge of the ship. In him I had every trust and confidence. He entered upon the enterprise with enthusiasm tempered with prudence and sound good sense. Thereafter the burden upon me was greatly lifted and his companionship was much appreciated, not only during the conduct of the exploration itself, but also in the trying months of preparation.

It would be an easy matter for me to dilate upon the splendid work accomplished by the " S.Y. Aurora," and the superb seamanship exhibited in the handling of the vessel during some of the hurricanes encountered, but that can be gathered by reading between the lines of Davis's text.

In conclusion, on behalf of the Land Parties of the Australasian Antarctic Expedition, I wish to express our gratitude to the intrepid captain of the " S.Y. Aurora," to his officers and to the crew for bringing the vessel with punctuality backwards and forwards through the ice-packed seas to our relief. But Davis and his men have done more than this, for by their oceanographic investigations they have carried out a piece of research which in scientific value is comparable with that accomplished ashore.

The publication of the book I await with interest, for so far, on account of the distance separating us, I have not yet had the opportunity of reading the text in detail.

No further introduction is needed from me for one who, by length of voyages undertaken, which fortunately have all ended successfully, must be regarded as the most experienced navigator of Antarctic waters.

D. MAWSON.

ADELAIDE,
August 15*th*, 1919.

CONTENTS

		PAGE
THE SOUTH LAND		1

CHAP		
I	THE SHIP AND HER STORY. EARLY YEARS IN NORTHERN WATERS	
II	THE S Y *Aurora*, 1911	10
III	CAPE TOWN TO HOBART	14
IV	FIRST ANTARCTIC VOYAGE HOBART TO MACQUARIE ISLAND	16
V	THE FIRST ANTARCTIC VOYAGE (*continued*) MACQUARIE ISLAND TO THE MAIN BASE	23
VI	A SYNOPSIS OF THE EARLIER VOYAGES ALONG THE COAST-LINE OF ANTARCTICA, IN THE AUSTRALIAN QUADRANT	33
VII	THE FIRST ANTARCTIC VOYAGE (*continued*) COMMONWEALTH BAY TO THE WESTERN BASE AND BACK TO HOBART	43
VIII	A WINTER CRUISE	57
IX	A WINTER CRUISE (*continued*)	69
X	THE SPRING CRUISE IN THE SUB-ANTARCTIC	75
XI	THE SECOND ANTARCTIC VOYAGE	82
XII	THE RELIEF OF THE WESTERN PARTY	97
XIII	THE MAWSON RELIEF FUND, 1913	106
XIV	THE THIRD ANTARCTIC VOYAGE	109
XV	THE LAST VISIT TO MACQUARIE ISLAND	116
XVI	FROM MACQUARIE ISLAND TO COMMONWEALTH BAY	121
XVII	AT COMMONWEALTH BAY	126
XVIII	WESTWARD HO !	133

CONTENTS

CHAP		PAGE
XIX	QUEEN MARY LAND	139
XX	THE HOMEWARD VOYAGE	147
XXI	THE COASTAL LINE OF PACK-ICE	151
XXII	THE D'URVILLE SEA	156
XXIII	THE *Vincennes* IN 1840, THE *Aurora* IN 1912	160
XXIV	L'AVENIR	166
APPENDIX I	LIST OF MEMBERS OF THE AUSTRALASIAN ANTARCTIC EXPEDITION	169
,, II	GLOSSARY OF ICE TERMS	171
,, III	THE ANTARCTIC REGIONS AS KNOWN IN 1914	175
,, IV	PLAN OF THE *Aurora*	177
INDEX		179

LIST OF ILLUSTRATIONS

	PAGE
John King Davis	*Frontispiece*
Douglas Mawson, D Sc , B E., Commander of the Australasian Antarctic Expedition, 1911–14 .	2
Professor T W Edgeworth David, C M G , D S.O , F R S	4
Professor Orme Masson, C B E , F R S , President of the Australasian Association for the Advancement of Science (1911)	6
The *Aurora*	8
Th *Aurora* fitting out in London, 1911	10
The Anto Hipon Stone and the London Stone	12
The Ward Room of the *Aurora*	14
"Running the Easting down " Mertz, Corner, Gray, Ninnis	14
Mertz up Aloft	14
The Motor Launch	14
The Officers of the *Aurora*	16
Charts of Macquarie Island, 1820–1914 (not on the same scale)	18
First Landing on Adelie Land	28
The Motor-boat Towing two Whale Boats—Landing Stores—Main Base —Commonwealth Bay, January, 1912	28
The *Aurora* at Anchor, Commonwealth Bay	30
Landing at the head of the Boat Harbour	30
Steaming through Loose Pack	46
The Wake of the Vessel through Loose Pack	46
Ice Barrier sighted February 8, 1912	49
Frank Wild, leader of the Western Party	52
The *Aurora* off Wild's Base	52
The Flying Fox used for hauling Stores from the Floe-Ice to the top of Cliff	52
Wild's Base on the Shackleton Shelf	53
Landing Stores at Wild's Base. Showing the top of the Cliff	53
The *Fram* at Hobart, Tasmania	55
The Monagasque Trawl Frame and Net .	58

LIST OF ILLUSTRATIONS

	PAGE
A Dynamometer	58
A Dredging Block	58
Th Monagasque Trawl	60
The Dredging Boom and part of the Wire Reel	60
Dredging Reel on board the *Aurora*	60
G F. Ainsworth, Leader of the Macquarie Island Party	64
The Macquarie Island Party outside the "Shack"	64
Mr. Waite outside the "Shack"	66
H M Mail arriving at Macquarie Island	66
The Landing at Caroline Cove, Macquarie Island	66
The *Aurora* at anchor off Erebus and Terror Coves, Auckland Islands	70
Depôt for Shipwrecked Mariners at Camp Cove, Auckland Islands	70
Rata Trees at Erebus Cove, Auckland Islands	72
Shipwrecked Mariners' Depôt, Camp Cove, Auckland Islands	72
The *Aurora* at Anchor, Carnley Harbour, Auckland Islands	72
The *Aurora* at anchor off Shoe Island, Auckland Islands	72
The Launch at Camp Cove, Auckland Islands	72
Sounding off Macquarie Island with Kelvin Machine	78
Main Base The Boat Harbour and Landing Place	86
Main Base, 1913	86
A Large Berg	90
Looking North in the Vicinity of Main Base	90
Large Floe with Seal	92
At "Aladdin's Cave"	92
Mackellar Islet with Ice Cap	92
The Decks of the *Aurora* after a Blizzard	94
Group of Party that remained for Second Year	96
Western Party taken just after Relief	102
Photo of "All Hands" taken after the Relief of Wild	102
A Heavy Floe	104
Penguins on the Sea Ice off Wild's Base	104
The *Aurora* in Dock, showing the Heavy Four-bladed Propeller	108
The *Aurora* in Dock at Williamstown, Victoria	108
Preparing the Whaler	122
The Biologist	122
The Cross erected on Cape Denison	124
Dr X Mertz	124
Lieut. B E S Ninnis	124
Inscription on the base of the Cross	124
Stillwell Island Off Adelie Land	126

LIST OF ILLUSTRATIONS

	PAGE
A Party Landing on the Mackellar Islands	126
Ice covered Rocks near Winter Quarters, Adelie Land	128
The Mertz Glacier Tongue	130
A Cave in the Ice-wall, Mertz Glacier Tongue	130
A Whale rising close to the Ship	134
A Whale Spouting	134
Whale just after Spouting	134
Placing the Blocks of Ice in Tank through which a steam coil passes	140
Watering Ship from the Floes	140
Pushing through Pack off Termination Tongue	142
A "Water Sky" is indicated by dark line on the Horizon	142
Edge of Tabular Iceberg off Shackleton Shelf	142
Off Termination Tongue	142
Loose Pack off Termination Tongue	142
Off Termination Barrier Tongue Ice Blink over Barrier	148
Close view of portion of Termination Barrier Tongue	148

SKETCH MAPS AND ILLUSTRATIONS IN TEXT

	PAGE
Sketch to illustrate the Retreat of the Greely Expedition	7
Track of the *Aurora*, first Antarctic Voyage	17
View of Hasselborough Bay	20
Caroline Cove, Macquarie Island	22
Bishop and Clerk Rocks	23
Appearance of barrier inside fringe of pack-ice 8 m., 3 1 12	25
Tracks of the *Aurora* and *Vincennes*	26
Track of the *Aurora* from noon 6 1 12	28
The Anchorage at Commonwealth Bay	29
The Balleny Islands	34
Côte Clarie from Durmont d'Urville's Chart	36
Track of the *Vincennes* from the Ross Tracing	38
Track of the *Vincennes* off Knox Land	38
Repulse Bay, position of *Vincennes* 17 2.40	39
Track of the *Gauss*, 1902–3	41
Track of the *Aurora* off Adelie Land	44
Track of the *Aurora* to 23 1 12	45
View of Ice Cliffs, Côte Clarie, from chart of d'Urville, 1840	45
Track from 23 1 12 to 24 1 12	47
Sketch of Course followed by the *Aurora*	49
Landing of Wild's Party, 1912.	51
Track of the *Aurora*, Sydney to Lyttleton	56
Plan illustrating the arrangements for Trawling	58
Olive Weights	59
Various charted positions of Royal Company Islands	61
Position assigned to Royal Company Island on our French Chart	63
The Auckland Group	68
Track of the *Aurora*, Spring Cruise, 1912	75
Track of the *Aurora* from Hobart to Macquarie Island	77
Deep Sea Soundings	78
Soundings in the Vicinity of the Mill Rise	79

SKETCH MAPS AND ILLUSTRATIONS IN TEXT

	PAGE
Depths north of Macquarie Island	81
Approximate Sketch of Soundings taken on the Mill Rise	81
Second Antarctic Voyage, track outward and homeward	82
Track Southward	83
Track Southward	84
Approaching Main Base 1912, approaching Main Base 1913	85
The Anchorage, Commonwealth Bay	86
5¼ mile Depôt	88
"Scanning the Coast"	93
Track of the *Aurora* 10 2 13	99
The Relief of Wild's Party, 1913	101
Track in the vicinity of Termination Ice-Tongue	104
Graph to illustrate Sea Temperature Section	111
Relief Voyage, Hobart to Adelaide, 1913–14	112
Section of Sea Floor between Tasmania and the Antarctic	113
Hobart to Macquarie Island	114
Section showing depths adjacent to Macquarie Island	117
Blake's Chart of Macquarie Island	120
Approaching the Main Base, 1913	123
Commonwealth Bay to Buchanan Bay	129
Track from December 31, 1913, to January 4, 1914	132
Track of the *Aurora* from 13 1 14 to 21 1 14	136
The Winter Quarters of the *Gauss*, 1902	138
Drygalski Island	139
Track of the *Aurora* off Queen Mary Land, January 19–January 27, 1914	140
Track of the *Aurora*	143
Track off Termination Ice-Tongue	144
Track of the *Aurora*	148
Pack off Ninnis Glacier-Tongue	151
Sketch showing the Northern Limit of the Coastal Pack in successive years off Termination Ice Tongue	152
The Northern Limit of Coastal Pack-ice	153
Sketch Map to illustrate Soundings	155
The d'Urville Sea	156
Tracks of the *Aurora* in the d'Urville Sea	157
Track of the *Vincennes*, 1840, and the *Aurora*, 1912	161
Position of the *Vincennes* 17 2 40	163
Position of the *Aurora* 8 2 12	163

(Photo, Swaine.

John King Davis

MAPS AND DIAGRAMS

	PAGE
The Australian Quadrant of Antarctica	1
Track of the *Aurora*, 1912	54
Queen Mary Land in 1914	144
North Eastern point of Termination Ice-Tongue	147
Track of the *Astrolabe*, 1840, and the *Aurora*, 1912 and 1914	154
Diagrams to illustrate notes on Wilkes's Chart	162
The Antarctic Regions as known in 1914	176
Regional map showing the Area covered by the Australasian Antarctic Expedition, 1911-14	*inside back cover*

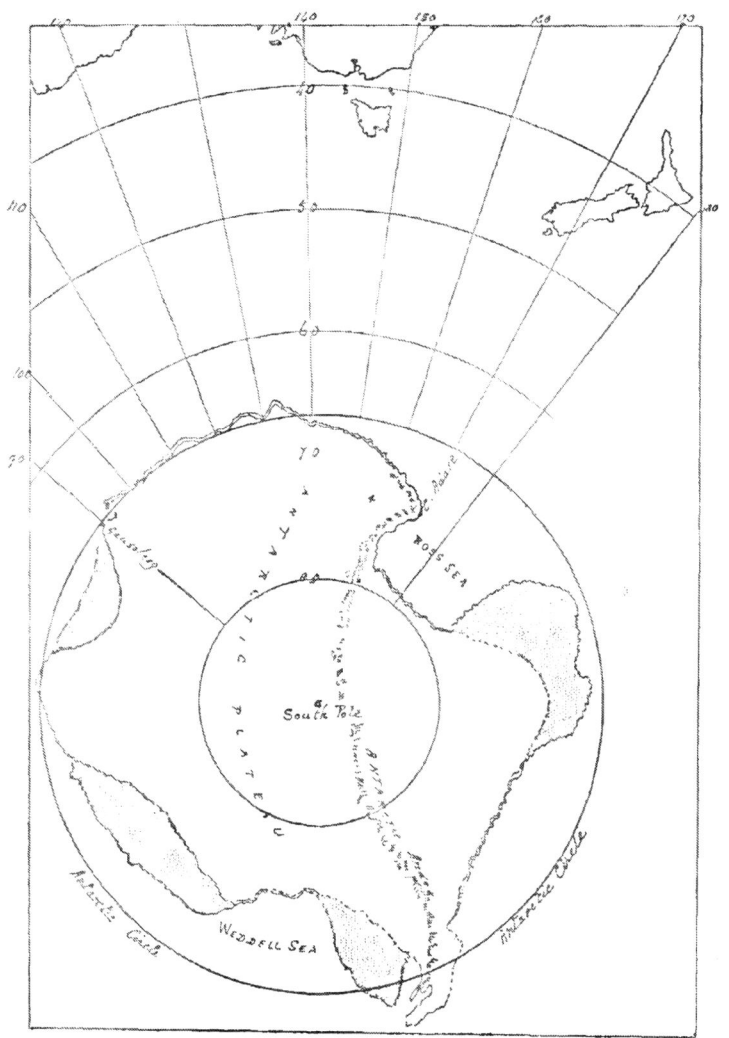

Sketch of Australian Quadrant, showing the section of the Antarctic coast line selected for examination by the Australasian Antarctic Expedition 1911. Also the projected outline of Antarctica from the Map of the South Polar Regions by Douglas Mawson, D.Sc., B.E.—*Geog. Journal* 1911 (June).

Portions shaded are probably Pack Ice. South Magnetic Pole, X.

THE SOUTH LAND

FROM the close of the fifteenth century, it was generally believed that an immense continent lay round the South Pole, extending towards the Equator in each of the great oceans. When Tasman sighted the west coast of New Zealand in December, 1642, he was under the impression that it was a promontory of the "Great South Land." Among the records preserved at the India Office, London, is a rough draught marked:

"*A draught of the South Land, lately discovered*, 1643."

Captain Cook in H.M.S. "Resolution," was the pioneer navigator (1773-75). Sailing from west to east, he crossed the Antarctic Circle for the first time in history, in 39° 35′ East, in search of this "South Land." In the course of his voyage he penetrated south of the Circle at three points, but, on each occasion, he found the land unapproachable, owing to a wide belt of impenetrable ice.

The observations of the earlier navigators, and the information more recently obtained by various expeditions, which have visited the Antarctic regions since the introduction of steam-driven vessels, have resulted in fixing the boundary of this "South Land" within certain limits, besides acquiring more definite knowledge of some inland parts of this ice-capped continent, now known as Antarctica. Scientists have roughly estimated its area as exceeding five million square miles; it nearly fills the space fringed by the Antarctic Circle and, just south of Cape Horn, the land extends beyond the Circle.

The accompanying diagram shows the probable outline of the Antarctic Continent, as deduced from the data available in 1911.

THE AUSTRALASIAN ANTARCTIC EXPEDITION (1911-14) selected as its zone of operations the coastal region in the Australian Quadrant from Cape Adare westward to Gaussberg, a distance of over 2,000 miles. Between these terminal points a landing had been made in 1840 (for a few hours) on a small rocky islet close to the ice-cliffs. The Expedition was organized and led by Dr. Mawson, an Englishman by birth, who went to Australia early in life, where he had a distinguished career at the University of Sydney He made special studies of geology and chemistry, finally graduating in Mining Engineering and in Natural Science. In 1905 he was appointed Lecturer in Geological Studies at the University of Adelaide.

Dr. Mawson is not only a distinguished scientist but also an experienced explorer. He was one of the most prominent members of the Shackleton Expedition (1907-09), being one of those who climbed Mount Erebus, and subsequently undertook the long journey to the Antarctic Pole (Magnetic).

On January 12th, 1911, Dr. Mawson placed his plans before the Australasian Association for the Advancement of Science. The Association warmly approved his project and voted a sum of £1,000 towards it. A committee of scientists was formed to assist in the necessary organization, and to approach their respective State Governments for funds. Professor Orme Masson, of Melbourne, Professor T. W. E. David, of Sydney, and Professor G. C. Henderson, of Adelaide, the three chosen members of the committee, proved enthusiastic supporters of this Australasian Expedition.

In the month of April of the same year, Dr. Mawson laid his plans before the Royal Geographical Society. After hearing the address, the President said: "As to the scientific results, I should like to take this opportunity of saying that the Council of this Society is so convinced of their value that we have decided to make a grant of £500 to this Expedition."

Generous contributions from individual donors in addition to the grants mentioned above enabled the leader to complete arrangements for the purchase of the "Aurora." There was some unavoidable delay in obtaining grants from the Imperial Government as well as from the Commonwealth and State Governments, but the matter of funds was settled satisfactorily before the time came for the Expedition to start from Hobart.

DOUGLAS MAWSON, D.Sc., B.E.
Commander of the Australasian Antarctic Expedition 1911-14.

The recruiting of the members was mainly from among graduates of the Universities of Australia and New Zealand. Those members who were not recruited in Australasia were in two instances experts; Dr. Mertz, the Swiss ski-champion in 1908, and Mr. Frank Wild, the veteran explorer of the " Far South." A complete list of the members of the Expedition is given in the appendix to this book.

The sympathy and support which his project had called forth in England and Scotland was much appreciated by Dr. Mawson. It should be specially mentioned that Dr. W. S. Bruce gave much valuable assistance in organizing the oceanographical work to be carried out in the Sub-Antarctic regions, while the Prince of Monaco took great interest in this part of the programme and contributed several articles of deep-sea equipment.

Before leaving England, Dr. Mawson stated to a representative of the *Morning Post* that he would like to emphasize the fact that the Australasian Expedition did not clash in any way with that of Captain Scott, that their fields of work were quite different and their mutual understanding of the most cordial character.

A new and practically unknown sphere of action had been selected, one nearer to Australia and more likely to yield results which would prove useful to Australians than the region to the west of Gaussberg or that to the east of the Ross Barrier. In any case the South Pole was *not* an objective.

Briefly the main objects of the Expedition were.—

1. To determine definitely the actual coast line of the continent between the terminal points.

2. To examine, so far as circumstances permitted, this coastal region from a scientific point of view.

The more important items of work required to carry out these objects would be.—

1. To land shore parties at intervals along the coast; each party being equipped with the necessaries for wintering during the polar night of 1912.

2. To make a complete survey of the coastal region, or as much as could be executed, by the despatch of sledging parties travelling along the coast east and west from each wintering station.

In order to correlate meteorological observations, a wireless station at the main Antarctic base and a wireless repeating station at Macquarie Island would be established. For the purpose of magnetic observations, an inland party would sledge from the main base towards the South Magnetic Pole, to complete some magnetic data yet wanting in that vicinity. The oceanographical work—sounding and trawling—would be carried out by the exploring vessel.

PROFESSOR T. W. EDGEWORTH DAVID, C.M.G., D.S.O., F.R.S.

CHAPTER I

THE SHIP AND HER STORY—EARLY YEARS IN NORTHERN WATERS

THE "Aurora" was built in 1876, and for the next ten years was one of the Dundee whaling fleet. These whalers were designed and built for navigation in northern seas. The hulls are of wood on account of its greater elasticity where pressed by ice, while the hardwood sheathing minimises the abrasion caused by contact with the jagged edges of the ice-floes. The cutwaters are protected with iron plates; steam power enabling the vessels to enter the pack ice, and occasionally to break through the obstructing belt.

Each vessel of the fleet used to make an annual voyage to the northern fishing grounds. From Dundee to St. John's the run is usually through open water. From the latter port the ship visits the sealing ground where some weeks are spent before returning to St. John's to prepare for a cruise to the whaling ground in Lancaster Sound or that off the island of Jan Mayen. Having secured a cargo of seal-skins and oil, the vessel returns to Dundee, to resume her routine work early in the following year.

During these annual voyages in the seventies, the seagoing qualities of the "Aurora" had been thoroughly tested. She had proved to be well adapted for ice-navigation, as well as an excellent sea-boat, when heavy weather was encountered in open waters.

The voyage of the year 1884 was a notable epoch in the earlier life of the "Aurora." In that year several Dundee whalers took part in the search for traces of the Greely Expedition which had occupied a circum-Polar Station at Discovery Harbour (81° 44′ N.) in August, 1881.

A short summary of the circumstances which led to the

dispatch of the Relief Expedition by the Navy Department of the United States in May, 1884, will explain the position which afforded the whalers an opportunity to display dash and energy to a remarkable extent.

THE AMERICAN STATION AT LADY FRANKLIN BAY, 81° 44′ N., 64° 15′ W.

The story of the Expedition may be epitomized as follows:—

Two years of success ending in August, 1883; and ten months of storm and stress ending in June, 1884.

During the first two years a well equipped party, twenty-four in number, under Lieut A. W. Greely, of the United States Army, landed at Discovery Harbour, Lady Franklin Bay, in August, 1881. They were provisioned for twenty-seven months. Important geographical discoveries were made by sledging parties, and a series of observations, meteorological and magnetic, was carried out during a period of two years.

The leader had been instructed, in the event of no relief ships being able to reach the station during the summer of 1882 or 1883, to leave the station and retreat southwards by boat until a relief ship was met or Littleton Island reached. A steam launch and three whale boats had been landed at the station in case retreat should be necessary.

On August 9th, 1883, as no relief ship had reached the station, Greely decided to retreat by boat. The conditions of the ice were most unfavourable, and the party met with great difficulties. The launch and two of the whale boats had to be abandoned. For thirty days they lived on the moving pack. At length the party landed at Baird Inlet, 300 miles south of Lady Franklin Bay, and proceeded to Cape Sabine. They had saved one whale boat, a sledge, one tent, all the records and most of the instruments, but the supply of food was only sufficient to give each man a reduced ration for a period of forty days. It was then as late in the season as October 26th.

Greely decided to winter near Cape Sabine, and when the ice conditions improved, to attempt to cross to Littleton Island. A hut was built; snow, stones, some canvas and a whale boat being the only materials available. Here the

PROFESSOR ORME MASSON, C.B.E., F.R.S.
President of the Australasian Association for the Advancement of Science (1911)

THE SHIP AND HER STORY

winter was spent, the men suffering acutely from cold and hunger. One man died in January, 1884, and six had perished before the end of April. As the hut had become untenable, a tent was pitched on higher ground. Altogether eleven deaths occurred during May and June of that fatal year.

Meanwhile the United States Navy Department had purchased two Dundee whalers, the "Thetis" and the "Bear," and a relief expedition under Commander Schley was organized. The vessels were manned by officers and men of the Navy. The "Bear," as pioneer ship, sailed for Cape York on May 4th. Nothing was known as to the locality where Greely and his men had passed the winter (1883-84).

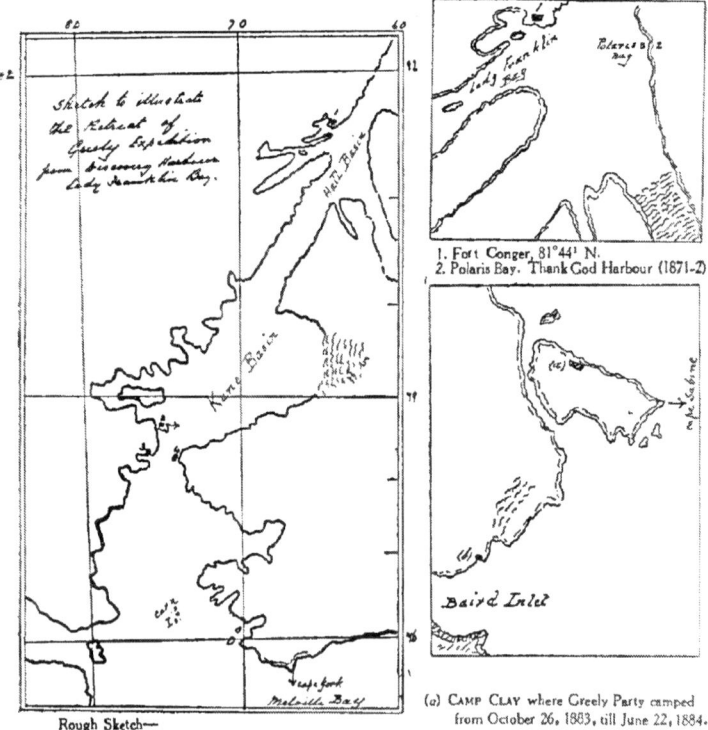

Rough Sketch—
1. Circum-polar Station, called Fort Conger.
2. Camp near Cape Sabine, called Camp Clay.
3. Eskimo Point.   4. Little'on Island.
5. Great Humboldt Glacier.

(a) CAMP CLAY where Greely Party camped from October 26, 1883, till June 22, 1884.

(b) Eskimo Point where Party landed, Sept. 1883.

A proclamation had been issued by the Navy Department offering a reward of 25,000 dollars to any ship, or to any person (not belonging to the Naval or Military Service of the United States), "who shall discover and rescue, or satisfactorily ascertain the fate of the Expedition under command of Lieut. Greely."

On the publication of this proclamation, four whalers sailed *direct* for Davis Strait from Dundee. The "Aurora," the "Arctic," and the "Wolf" were then fitting out at St. John's for the annual whaling cruise. Their preparation was hurried forward to enable the vessels to take part in the search. It was arranged that if any whaler came on traces of the expedition, the information would be reported to one of the relief ships with the least possible delay. The whalers would have a friendly but exciting race across Melville Bay, and the first ship to reach Cape York might possibly hear something from native sources, or even meet some members of the expedition.

On May 8th the "Aurora" left St. John's fully equipped for a whaling cruise. She was manned by a complement of sixty-five, under the command of Captain Fairweather. Mr. Lindsay, F.R G.S., was the medical officer on board, and his narrative of the various incidents which occurred during this cruise is graphically told and well worth reading.*

On June 18th the "Aurora," "Bear," "Thetis" and "Wolf" cleared the pack with open water ahead, close to Cape York. The "Aurora" was leading until close to the shore, where she was passed by the "Bear." The latter touched ice four miles from Cape York, the "Aurora" being about a mile in the rear; the other ships following in her wake. No information was obtained at Cape York, so the vessels proceeded northwards to examine a supply depôt at Cary Island.

On June 21st the "Aurora" and the "Bear" reached Cary Island: the Depôt was examined, but Greely had not been there. The whalers were now about to part from the relief ships and proceed to Lancaster Sound.

The "Bear" left Cary Island about midnight, and within

* *A Cruise to the Arctic in the Whaler "Aurora,"* 1884, by Lindsay, D.H. Published in 1911.

THE "AURORA." [Photo, A. H. Ninnis.

THE SHIP AND HER STORY

twenty-four hours the seven survivors of the expedition had been rescued at Cape Sabine.

How the "Aurora" was nipped in the ice between Cape York and Cary Island is described in vivid language by Mr. Lindsay. For some hours the ship was in danger of being crushed by the heavy pressure, then the ice opened by degrees and the vessel was let down into the water without having sustained serious damage

Altogether the "Aurora" had thirty-four years' service in northern waters before being purchased by Dr. Mawson as the exploring vessel of the Australasian Antarctic Expedition. In the following chapters I have tried to set down in simple language various incidents which occurred during the voyages of the vessel in the Antarctic and Sub-Antarctic regions of the Australian Quadrant throughout a period of more than two years.

A sailor cannot but regard with affection the good ship which has carried her living freight in perfect safety over many miles of tempestuous ocean, and through the berg-strewn seas of the Southern Ocean. After five voyages the "Aurora,"* though somewhat scarred and weather-beaten, was still sound and serviceable in February, 1914.

* Since these lines were written the "Aurora" mysteriously disappeared while on a passage across the South Pacific. The ship sailed from Sydney, New South Wales, on June 20th, 1917, bound for Iquique, Chili, but failed to arrive at her destination On her last fateful passage she was carrying coal, and was finally posted as "missing" at Lloyd's on 2nd January, 1918.

CHAPTER II

THE S.Y. "AURORA," 1911

In June, 1911, the vessel arrived in the Thames, where, after being docked, she was thoroughly overhauled and refitted in the South West India Docks under my supervision. She was entering on a new phase of her career as an exploring vessel on special service. Various alterations were necessary to meet the requirements of the expedition. New and improved accommodation had to be provided for the convenience and comfort of the explorers while being conveyed to, and brought home from, the wintering stations in the Antarctic.

The "Aurora" was fitted with a new foremast and the rig was altered to that of a barquentine; sails being carried for use in the region of westerly winds as well as to provide means of escape in case of injury to the propeller while working through the pack-ice. In 1881 the ship had been fitted with a two-bladed propeller, but she now carried a four-bladed, built propeller capable of withstanding very hard knocks. The double-wheel by which the vessel had been steered from right aft was replaced by gear erected on the bridge. Two laboratories were provided on deck for the use of scientists while investigating the mysteries of deep-sea life as revealed from time to time by the contents of the trawl. The topgallant forecastle was used as a store for various articles of equipment, while the deck below was specially fitted up as quarters for the crew. The wardroom aft was altered to accommodate twenty-five members when necessary, and special attention was given to ventilation throughout the ship. Two sounding machines were fitted on deck. No. 1 Lucas was fixed up on the port side of the forecastle head, the wire being wound in by means of a belt worked by a small horizontal engine. This engine had been constructed for the "Scotia" (1902),

THE "AURORA" FITTING OUT IN LONDON, 1911

and was very kindly lent to the Expedition by Dr. W. S. Bruce. (The sounding and trawling gear will be described in more detail in a later chapter.)

A motor-launch, masts for wireless stations and various articles of equipment for the landing parties were under construction in Sydney, and were to be embarked when the "Aurora" reached Hobart.

After some weeks of preparation for the work before us in southern waters, we embarked some three thousand cases of stores, besides forty-eight Greenland dogs, which had been procured through the good offices of the Danish Government, for sledging work. The animals varied in size, colour and general appearance; some were about the size of a collie, while others were much larger.

On July 28th we dropped down the river to Sheerness shortly before midnight The dogs did not appear to relish the idea of another sea voyage as they howled and barked in a manner expressive of strong displeasure, commencing a continuous "Danish" concert; an indescribable discord, which aroused the sleepers on board several river-craft at anchor, and called forth from the awakened individuals remonstrances couched in nautical language which will not bear translation.

Certain instruments lent by the Admiralty were taken on board before proceeding down Channel. We had arranged to take five hundred tons of fuel (in 25 lb. blocks) from Cardiff. This fuel was easily stowed and proved satisfactory for steaming purposes.

On August 4th we sailed from Cardiff for Table Bay—the ship's complement being twenty-one. Two members of the Expedition—Lieut. Ninnis and Dr. Mertz—sailed with us.

14.8 11. Madeira was sighted this morning There is a strong trade-wind. The ship has been doing six knots on a consumption of 4 7 tons of fuel a day. With a strong, fair wind we can average seven-and-a-half knots. The deck has been cleared and the ship made presentable after the bustle of leaving port. The dogs have more room, but they do not enjoy the sultry weather.

28 8.11. We crossed the line to-day in longitude 21°W., twenty-four days out from Cardiff. The dogs had a scamper round the deck and evidently enjoyed the exercise.

24.9.11. Capetown was reached at 9 a.m., thus completing the first stage of our voyage to Hobart in forty-nine days. We shall take fifty-four tons of the best Natal steam coal from here. The price is reasonable—22s. 6d. per ton—and the Chief Engineer considers it fairly good steaming coal.

We have had a fine-weather voyage so far, but perhaps Mertz and Ninnis found it somewhat monotonous. However, we must expect a breeze or two before reaching Tasmania, and the dogs may suffer some inconvenience in heavy weather. Before we sail I shall get some fish to tempt the canine appetite as a substitute for the regular ration of dog biscuit.

* * * * *

Since the beginning of the seventeenth century many vessels other than Portuguese called at Table Bay to obtain oxen and sheep from the natives. The captains of vessels outward bound used to leave letters under a large flat stone and carve on the upper surface :

"Hereunder look for letters."

These documents would be taken in charge by the captain of a homeward bound vessel, who, in his turn, would leave papers reporting his arrival at Table Bay for transmission to India or Batavia. Most of the inscribed stones which have been recovered have been found near the old landing place at the lower end of Adderley Street, when digging foundations for modern buildings.

Each of these " Post Office Stones " bears a date, the name of a ship and that of her captain. The stone which bears the oldest date in English records was discovered embedded in one of the walls of the Castle; the dimensions being about 82 cm. by 80 cm.

The inscription reads :—

"Anto Hipon Ma(ster) of the Hector Bound Home January 1605.
Anto Hippon Ma(ster) of the Dragon 28 December 1607"

Anthony Hippon had put in to Table Bay as Master of the " Dragon " in 1607. He found his first graven stone, and added to it the date of his second visit.

Another stone is on view in the vestibule of the General Post Office, Adderley Street. The second or lower inscription appears to be a Dutch one.

THE ANTO HIPON STONE.

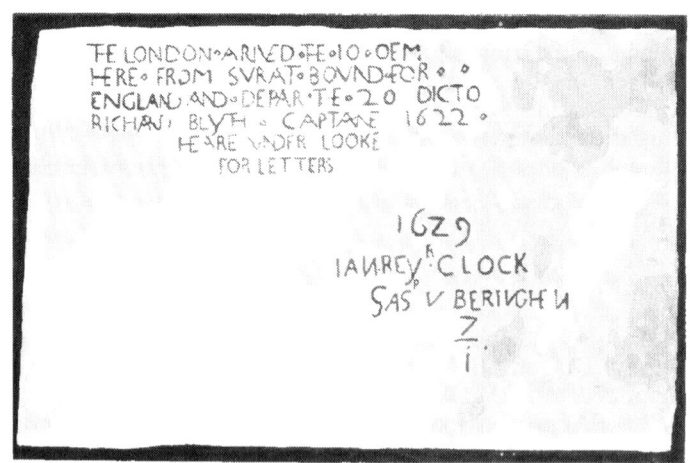

THE LONDON STONE.

The upper inscription reads :—

The London arrived the 10 of M(arch) here from Surat Bound for England and Depar(ted) the 20 Dicto 1622

Richard Blyth Captain
Hereunder looke for letters.

Below this :—

1629. Jan. Reyr. Clock.
Gasp. V Beringhen
$\frac{7}{1}$

CHAPTER III

CAPE TOWN TO HOBART

Our shipping agents had given us every assistance, and we were able to resume the voyage on September 27th.

28.9.11. We crossed the Agulhas Bank to-day. A sounding was taken in 110 fathoms on sand and small stones. We lost two of our dogs from what appeared to be distemper. A S.E. breeze has sprung up rather suddenly and caused quite a sea.

3.10.11. We got a boat out to-day and several photos of the ship were taken by Mertz and Ninnis The former was very fond of climbing aloft and taking views of the deck from various points. At first we were inclined to suggest caution when climbing about the rigging, but we very soon realized that he was quite as agile as any man on board.

8.10.11. We are making good progress under top-sails, favoured by a fresh N.W. breeze. The "Aurora" steers well, but with a following sea she can kick hard. Run, 183 miles.

11.10 11. The last twenty-four hours have been our first experience of the ship in a really heavy sea. She is an excellent sea boat, and our run has been 223 miles.

This has beaten the "Nimrod's" record by eight miles.*

12.10.11. W.S.W. gale with snow squalls.

13.10.11. W.N.W. gale. Squalls with rain. Some squalls lasting for half-an-hour were violent, the wind blowing with hurricane force.

16.10.11. After a week's gale, the wind has moderated to-day. This gale has been one of the worst I have ever experienced. Our damage was confined to the loss of a few planks out of the bulwarks. We have shipped the old double wheel aft in case anything should happen to the steering

* The "Nimrod," while on a voyage from Sydney to Cape Horn via Macquarie Island, made a run of 215 miles under steam and sail on June 1st, 1909. Latitude 57° S., longitude 167° E. (approximate)

[Photo, Mertz.
THE WARD ROOM OF THE "AURORA."

Mertz. Corner. Gray. Nunnis. [Photo, Gillies
"RUNNING THE EASTING DOWN."

[Photo, Gray
MERTZ UP ALOFT

[Photo, Davis.
THE MOTOR LAUNCH.

gear on the bridge. The dogs are getting a respite, but deck passengers have a very bad time in such inclement weather.

20.10.11. With the exception of October 16th we have had very bad weather for ten days, however we are making good progress towards our destination. At noon to-day we were 2,600 miles from Needle Rock, Tasmania. All through this heavy weather Mertz and Ninnis were always ready to turn out and lend a hand, and their assistance was most welcome.

During the next fortnight the weather improved considerably to the enjoyment of all on board. We were able to get some painting done and the " Aurora " had a " wash and brush up " before entering port.

Mertz took a flashlight photograph of the ward room.

4.11.11. We reached Hobart to-day. The dogs have been transferred to the quarantine station until the vessel is ready to sail south. We shall have a busy time until the end of the month completing the arrangements for the voyage to Antarctica.

We found a variety of equipment awaiting shipment, two complete wireless outfits, materials for three huts—one to be erected at each wintering station—an air-tractor sledge and a motor launch. These represented two of the more bulky packages.

The motor launch was driven by a petrol engine of 8 H.P. On the trial trip in Sydney Harbour a speed of seven knots had been attained. It had been specially strengthened for working in icy latitudes where a boat of this kind is liable to receive hard knocks from stray pieces of floe. It should be of great service in landing stores at the southern bases.

Room was found on the skids and forecastle head for the air-tractor sledge, known to members of the Expedition as the " Grasshopper." This was packed in a case about thirty feet long, and the total weight was about two tons—rather an awkward package to handle.

By the end of November we were freighted with some 150 tons of stores, and 386 tons of coal. The " Aurora " was now ready to start on her first voyage to the South as the " Aurora Australis."

CHAPTER IV

THE FIRST ANTARCTIC VOYAGE

HOBART TO MACQUARIE ISLAND

> 'Tis easy for our Naval Board,
> 'Tis easy for our Civic Lord
> Of London and of ease,
>   That lies in ninety feet of down
>   With fur on his nocturnal gown
> To talk of Frozen Seas!
>   *Thomas Hood,* " Ode to Captain Parry "

ON December 2nd, 1911, we left the wharf at 4 p.m., and as the ship drew away, ringing cheers wished God-speed to the first Antarctic Expedition under Australian leadership. H.M.S. " Fantome " and several steamers had dressed ship for the occasion. Dr. Mawson and thirteen members of the Expedition had joined the " Aurora," so with our ship's company of twenty-five all told, a total was reached of thirty-nine. The S.S. " Toroa " (Captain Holyman) had been chartered to act as tender to the " Aurora," to convey the remaining members of the landing party, with some cargo, as far as Macquarie Island.

We steamed down the river to the Nubeena Quarantine Station, where the dogs were taken on board. The animals looked all the better for their visit to the Station, where they had enjoyed plenty of exercise with a generous diet. The pack appeared in fine condition as one after another the dogs were tethered round the bulwarks.

Before leaving the shelter of the land we had to see that the deck cargo was securely lashed, as earlier in the day we had been warned by Mr. Hunt, the Commonwealth Meteorologist, to " expect fresh south-westerly winds." After leaving Cape Frederick Henry the forecast proved correct.

About 8.45 p.m. the following message was signalled

J. H. BLAIR.
Chief Officer (final voyage).

F. J. GILLIES.
Chief Engineer.

P. GRAY.
Navigating Officer.

C. P. DE LA MOTTE.
3rd Officer.

THE OFFICERS OF THE "AURORA."

by the Morse lamp to the Station at Mount Nelson: "Everything snug on board; ready for anything. Good-bye.—Hannam."*

We came out of Storm Bay into a fresh south-westerly wind and heavy sea. We had a heavy roll all day on the 3rd, although the wind was well abaft the beam. Next day there

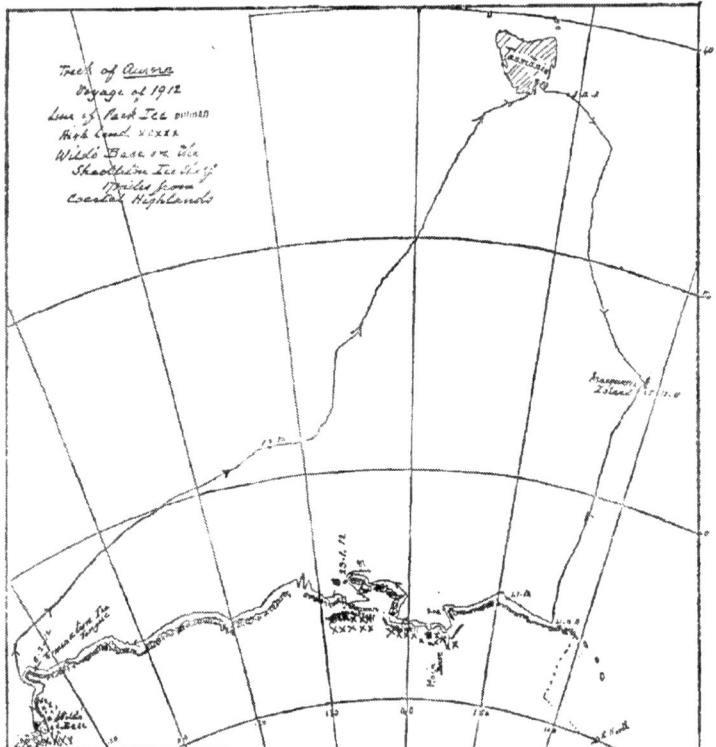

Track of the "Aurora" first Antarctic voyage.

was a short interval of moderate weather, but the glass had fallen rapidly, and at sunset a strong southerly gale set in with a very heavy sea. Very trying weather prevailed until the 9th, and the dogs, as deck passengers, once more had a bad time. One heavy sea smashed the case of the "Grasshopper."

* Hannam was a member of the landing party.

During this spell of tempestuous weather a great deal of water came *on-board*, but the consequent damage was light. Part of the bridge was the only loss *over-board*.

December 10th.—A fine day was a welcome change to all hands. From good observations, our position at noon was 106 miles from the north end of Macquarie Island.

11.12.11. A peculiar compass disturbance was observed at 1 a m. this morning. Both standard and steering compasses swung through five points for about four minutes. The island was sighted at 4 a.m. As the wind was easterly, we steamed down the west coast towards Caroline Cove, which was reached at 11 a.m. A sounding, taken about a mile from the shore, gave *no bottom* at 250 fathoms.

We towed a boat with a landing party through the bay to the north of the entrance to the Cove. The boat cast off when close in-shore and, shortly afterwards, the ship struck a submerged rock, carrying about 12 ft. of water, while the soundings taken fore and aft at the time gave 13 fathoms. As the vessel was moving very slowly, no damage was done. After the return of the shore party, we moved out and lay off the Cove until daylight.

As Macquarie Island will be frequently referred to in the course of my narrative, a few notes on the locality are introduced here.

The island is about 545 miles from the southern extremity of New Zealand and about 850 miles from Hobart. Captain Hasselborough of the brig "Perseverance," landed there in 1809, but, as he saw the remains of a wreck on the coast, it may have been visited by some navigator at an earlier date

The main island has a length of over 20 miles and an average breadth of about 3 miles. Reefs and rocky islands are numerous round its shores. Lying in a north-easterly and south-westerly direction it forms a breakwater, as it stands exposed to the full force of the prevailing westerly winds Two rocky groups known as the "Judge and Clerk," and the "Bishop and Clerk" rise from the submarine ridge (a continuation of the main island), to the N E and S W., a distance of 8 miles and 19 miles respectively.

Although the only vegetation is tussock grass and Kerguelen cabbage, there is a prolific animal life of birds and sea-elephants The fur seal is seen occasionally, but the ruthless slaughter of these animals in the early years of the nineteenth century almost exterminated them

My first visit to the island in 1909 led to the discovery of a solitary sailor living close to Nugget Point He had served in the Royal Navy for over twenty years and had come to Macquarie Island as one of the crew of a small schooner which visited the place once a year to collect sea-elephant oil When the ship was ready to sail, "Crusoe II" announced his intention of remaining on the

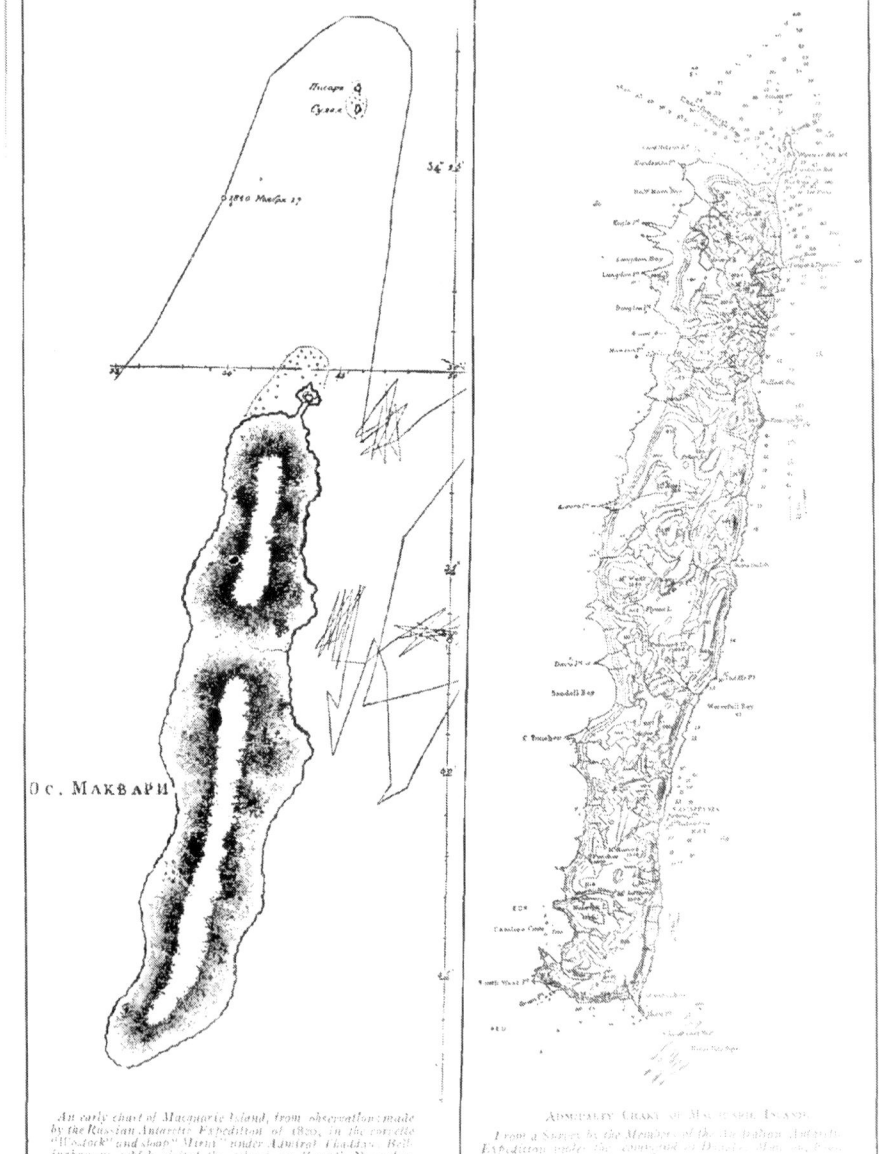

An early chart of Macquarie Island, from observations made by the Russian Antarctic Expedition of 1820, in the corvette "Vostock" and sloop "Mirni" under Admiral Thaddeus Bellingshausen, which visited the island on the 17th November 1820, while on a voyage from Sydney, N.S.W., to the Antarctic.

ADMIRALTY CHART OF MACQUARIE ISLAND.
From a Survey by the Members of the Australian Antarctic Expedition under the command of Douglas Mawson, B.Sc., D.Sc., B.E., S.Y. "Aurora", 1911.

island. As he was deaf to all persuasion, stores had been landed for his use. His hut of two rooms was warm and cosy. Each room had a stove, the coal for which he had to carry from the depôt—a distance of 4 miles.

He was not living alone on the island because he liked solitude—he was actually making money. As he explained to me, " There are sea-elephants and penguins and sea-lions. They mean oil, and oil means money." So he was collecting oil, of which he had already secured a valuable store which was housed in a galvanized iron building near his hut. This was known as the " Digester House," and it contained various appliances for converting the blubber into oil. He was a quaint character, and seemed thoroughly to enjoy life on this foggy and wind-swept island. When I proposed that he should return to civilisation on board the " Nimrod," he replied, " Why should I ? I'm happy enough here and have all I want. I'm glad to have seen you, but I don't want to leave the island." *

12.12.11. We steamed up the west coast and rounded the north end of the island and anchored off North-East Bay in ten fathoms. The wreck of a schooner was seen about four miles from the spot where the " Jessie Nicholl " went ashore in 1910.

A few hours later we heard that this was the " Clyde," owned by Mr. Hatch, the lessee of Macquarie Island. The schooner had sailed from Wellington with stores and coal for the sealers working on the island. She had anchored on November 13th. An easterly gale set in that night, and early on the 14th the cable parted and the vessel was driven on the rocks, becoming a total wreck. The crew reached the shore and obtained food and shelter from the sealers who were stationed about four miles from the scene of the wreck.

After we had dropped anchor we could see several men moving about on the shore, who signalled to the " Aurora " to come round to Hasselborough Bay, as the surf was too heavy to land on the eastern side of the peninsula. We steamed round Elliott's Reef to the western side and entered Hasselborough Bay, where we were met by Mr. Bauer,† who had put off in a boat manned by the sealers. He told us

* The history of Macquarie Island and of its former occupation by sealing gangs is given, at some length, in a book called *Murihiku*, by Robert MacNab (Whitecombe & Tombs, Ltd., 1909).

† Mr. Bauer represented the lessee at Macquarie Island. The men were working for the lessee, who finds the capital and pays the men at the rate of £4 per ton for the oil, the men paying for the food supplied. This oil is obtained from the sea-elephants, these animals being very numerous on the eastern coast.

20 WITH THE "AURORA" IN THE ANTARCTIC

that the wind had been easterly for nearly a month; a most unusual occurrence in that locality. At noon we dropped anchor in 15 fathoms.

The dogs were landed in charge of Lieut. Ninnis. Preparations were made to tow the wireless masts to the landing place, and at 7.30 p.m. a boat left the ship with the masts in

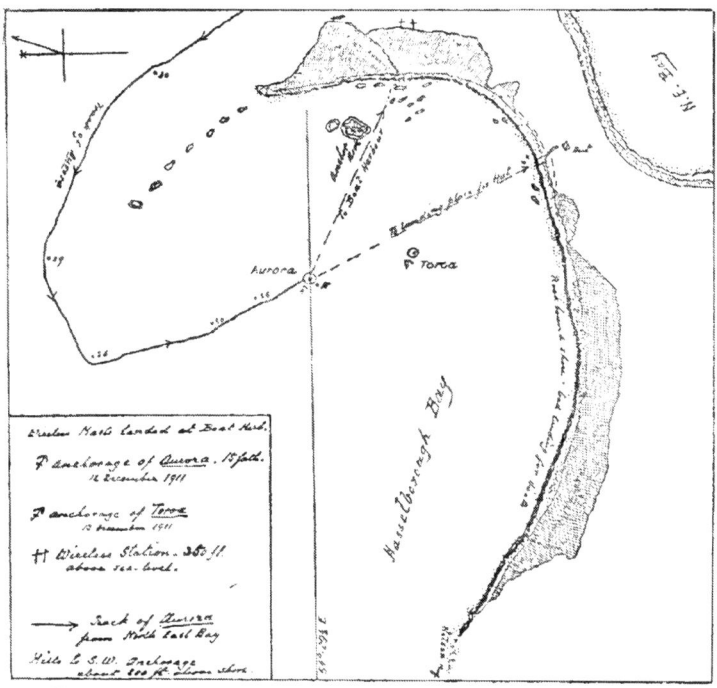

View of Hasselborough Bay, from sketch by Mr. Gray, second officer of "Aurora."

tow. The boat returned at 10.45 p.m., having accomplished its mission.

13.12.11. Our launch and boats were busy all the forenoon landing stores and wireless gear for the use of the party who will remain at this Station (the first of the three stations occupied by members of the Expedition in 1912). Mr. Ainsworth was in charge of the station, and his colleagues were Messrs. Blake, Hamilton, Sawyer, and Sandell.

About noon the smoke of a steamer was visible on the horizon. This proved to be our tender, the "Toroa" who, on her arrival, anchored about a mile from the "Aurora." Captain Holyman had a most fortunate passage of four and a half days from Hobart. The "Toroa" brought seventeen members of the Expedition, a new whale boat, ninety tons of coal and seventy-one cases of benzine to be transferred to the "Aurora." There were also fifty-five sheep, some of which were taken south; the rest being left on the Island. About ten tons of stores were landed for the use of the Macquarie Island Party.

16.12.11. Our launch conveyed the first "homeward mail" to the "Toroa," and this vessel sailed for Hobart at 10 p m Her passenger list included Mr. Eitel (secretary to the Expedition), Mr. Bauer, and five of his sealers, with the officers and men of the ill-fated "Clyde." Her freight included about two-thirds of the oil which was ready for shipment when the "Clyde" was wrecked.

The return voyage of the "Toroa" occupied five and a half days.

22.12.11. The work at the signal station was very nearly finished. Dr Mawson and his party came on board this evening. The whale boat brought off fifteen carcasses of mutton as a contribution to our Christmas larder.

23.12.11. The dogs were brought back to the ship, and the launch was hoisted out of the water and stowed on the skids, the weather being squally, with the glass falling slightly. As the surf in North-East Bay was too heavy to send the boat ashore for water, I decided to water ship at Caroline Cove.

24.12.11. We left the anchorage this forenoon and reached Caroline Cove about 2 p.m. Proceeding with caution into the Cove, we anchored in ten-and-a-half fathoms. A stream anchor was put out on the starboard quarter, as there was little room for the ship to swing. Watering ship was effected by towing two 100-gallon casks from the ship to the shore, where they were filled at a small creek, and then floated back to the vessel. At 8.30 p.m. we had a fair supply of good water on board, and work ceased for the night.

25.12.11. About 3 a.m. a sudden squall caused the ship to drag her anchor, and she touched the ground aft. A few revolutions of the engines relieved the situation and no damage was sustained, beyond the loss of the stream anchor. After this narrow escape, we moved out into deeper water, and I decided to put to sea as soon as possible, and to forego the pleasure of spending Christmas Day in Caroline Cove as had been intended. At 6 a.m. we were ready for sea, and a course was set for Antarctica.

A great deal of work has been done at Macquarie Island during the last fortnight. We were fortunate in having no westerly winds to contend against or to interfere with transshipment and landing operations. The first Station of the A.A.E. has been occupied without any serious accident, and the Island Party, under Mr. Ainsworth, has a fine field of operations.

Caroline Cove, Macquarie Island.

CHAPTER V

THE FIRST ANTARCTIC VOYAGE

From Macquarie Island to the Main Base

"Let us probe the silent places, let us seek what luck betide us!"
Robert W. Service.

Christmas Day, 1911. At 10 a.m. the rocky islets known as the "Bishop and Clerk" were abeam.

A. J. H.

Our voyage is beginning under favourable conditions, with a fair wind, clear weather and all sail set. We are now a party of fifty, all told, and it will take a day or two to settle down. The early hours of Christmas Day have witnessed rather a narrow escape, but we are now in full progress southward and have enjoyed an excellent Christmas dinner.

The members of the landing parties have been told off in watches. They assist with the pumping, bracing yards, making or furling sail, steering, hoisting boats in and out, and other miscellaneous duties. I think most of them rather enjoy keeping watch.

Later on in the Antarctic, when it was really necessary to call upon them for assistance, they proved a welcome and valuable addition to the strength of our small crew.

26.12.11. Our position at noon was latitude $57° 15\frac{1}{2}'$ S., longitude $157° 25'$ E. The run was disappointing (128 miles), but as the ship has not been docked since she left London, marine growths below the waterline must retard her progress.

We shall follow approximately the 157th meridian on a southerly course, until land is sighted, or until we are brought up by the ice. Land was reported by the "Terra Nova" about 700 miles to the south of our position at noon to-day.

The weather was remarkably fine during the day. Towards evening fog came on, but the sea continued smooth and the temperature was 45° F. This fine weather has greatly facilitated the repair of the launch which received a nasty knock just before we left Hasselborough Bay. It rendered excellent service at Macquarie Island, and there is plenty of work in store for it on reaching the next station where Dr. Mawson will establish winter quarters.

29.12.11. Dull, foggy weather with smooth sea has prevailed for the last forty-eight hours. No sights for position were obtainable to-day. At 2 p.m. the first Antarctic petrel was seen. Two hours later we encountered the first ice—several large bergs, one of which, about three-quarters of a mile long and some 70 ft. above the water, lay right across our course. From its appearance it might have broken away from the continental barrier on the previous day. The "blues" and "greens" were absolutely beyond description; at 6.20 p.m a long line of brash ice running north and south was passed to the westward. The temperature at midnight was 33·5° F.

30.12.11. Weather conditions remained unchanged Early this morning we passed close to what must have been a large berg, as the noise of the sea breaking on it sounded like distant thunder, but nothing could be *seen* owing to the fog. About midnight we were brought up by heavy pack which extended all round the southern horizon, so the course was altered to the west.

31.12.11. The ship is steering to the westward along the edge of the pack and passing many bergs and drift ice This morning a sea-leopard was shot on the floe and the carcase hauled on board. The skin measured 8 feet 8 inches in length and 4 feet 11 inches in girth. The carcase will be much appreciated by the dogs as a New Year's gift.

It is not often that one sees "the New Year in," in nearly broad daylight, and the members of the shore party mustered on deck just before the last "eight bells" of 1911. A few minutes later a group was photographed on the bridge, but, as the sun was about a degree below the horizon, the light was rather poor.

New Year's Day, 1912. Latitude 65° 18′ S., longitude 151° 50′ E.

All day steaming along the edge of pack, which seems interminable and of a solid nature with large earth-stained bergs marooned in it. Our course was approximately W.S.W.* Towards evening we were able to pass through some light floe ice which fringed the dense pack to the south. The weather continues dull and cloudy.

2.1.12. A long day was spent in trying to get south, but seldom we made better than west. At 8 p.m. the pack edge turned in a northerly direction, and we had to follow this course, though very reluctantly. Several sheep were killed to-

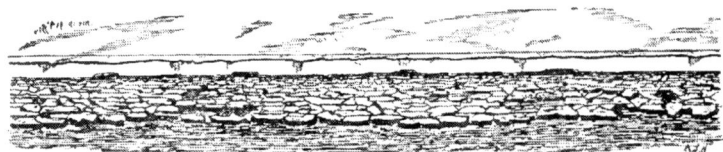

Appearance of barrier inside fringe of pack ice, 8 a.m., 3. 1. 12. A. J. H.

day, and the rigging is now decorated with carcases of mutton.

There has been a heavy north-easterly swell for the last twelve hours. The first giant petrel was seen to-day.

3.1.12. At 4 a.m. the direction of the edge of the pack changed to S.W. (approximately). A long barrier (or ice wall), about 60 ft. high, running E.N.E. and W.S.W., could be seen about a mile inside the pack ice. We followed this to its western end, about 5 miles distant. Open water could then be seen extending to the southward inside a narrow fringe of pack. On pushing through this fringe we followed the direction of the barrier (here about 70 ft. high), to the S.E. The way South seemed open at last!

8 a.m. We were now in open water just under the western face of the barrier running south. This course was followed for about fifteen miles, when the face trended to S.E., and course was altered to that direction.

Noon. Observations gave our position at noon as latitude

* Courses and bearings are *true* unless otherwise indicated.

26 WITH THE "AURORA" IN THE ANTARCTIC

65° 46′ S., longitude 143° 21′ E. During the morning the first Weddell seals were seen. The glass was falling and the wind increasing from S.S.E., so I decided to wait under the lee of the northern end of the barrier until the weather improved.

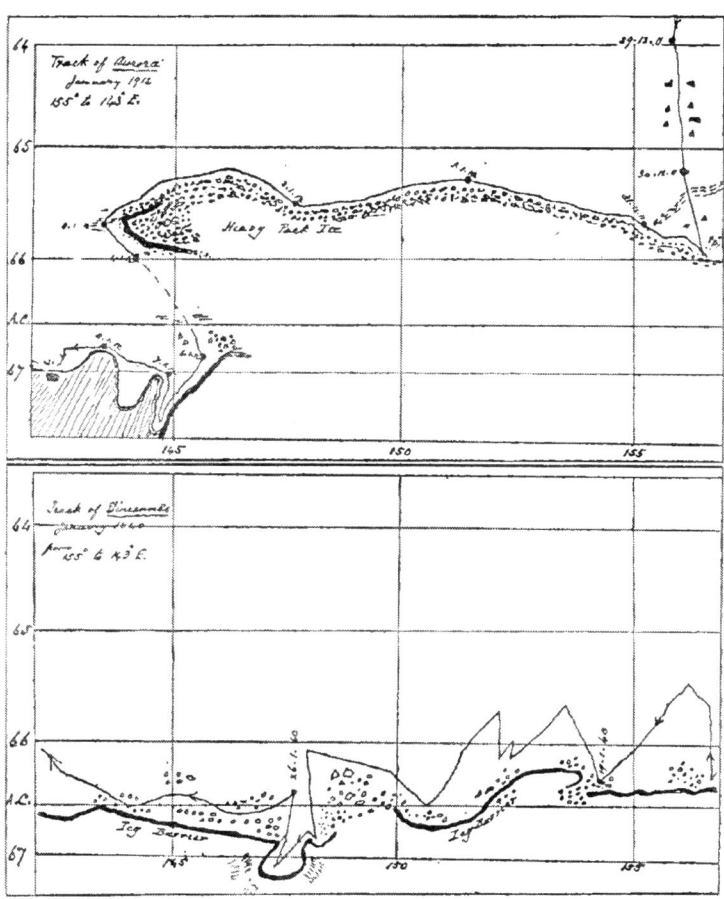

Tracks of "Aurora" and "Vincennes."

3 p.m. We met a large ice-island about 80 ft. high. The water being fairly smooth, a sounding was taken off this island in 208 fathoms on glacial mud.

A blizzard commenced shortly afterwards and continued

with very brief intervals until late in the afternoon of January 5th. During the blizzard we sheltered under the lee of the barrier, steaming slowly to and fro.

We were all rather puzzled to account for this huge mass of ice-barrier (or ice-tongue) which seems to prevent the pack from moving West

The track of Wilkes in 1840 is shown on the chart 30 miles further south. Possibly this barrier did not exist at that period.

The sketch shows the track of the "Vincennes" along the "Icy Barrier," and the track of "Aurora."

5.1.12. Our position at noon was in latitude 65° 41′ S., longitude 144° E. This was a little to the north of the locality where we had met the ice island on January 3. A sounding was taken in 230 fathoms, but the wire parted while heaving in. Later in the day we passed the ice-island (mentioned above), which had drifted some twelve miles to the N.W from its former position.

About 6 p.m. we were able to leave shelter and steer S.E. At 10 p.m. the line of ice-cliff trended to the east and was soon lost to sight. A few hours later we met some bay-ice, but no pack.

6.1.12. This morning the sky was heavily overcast and no sights for longitude could be taken. An observation at noon placed us in latitude 66° 37′ S., showing that we had crossed the Circle, and were approaching the coast-line.

An ice-cliff is showing right ahead, extending to the eastward and disappearing towards S.S W. It appears to be much higher than the cliff which afforded us such good shelter during the recent blizzard.

About 7 p.m. land was sighted to the S.S W We followed the ice-cliff to the head of a bay, and about 10 p.m. three islands were observed off the western end of this bay. The engines were stopped so as to allow the ship to lie off the outer island until the weather cleared, as snow was falling heavily.

7.1.12. The weather cleared about 1 p.m , and we were able to follow the line of coast to the westward at slow speed, passing numerous bergs and small islands. Soundings at intervals varied from 25 fathoms to "no bottom." At 10.45 p.m. we lay-to off the coast for the night.

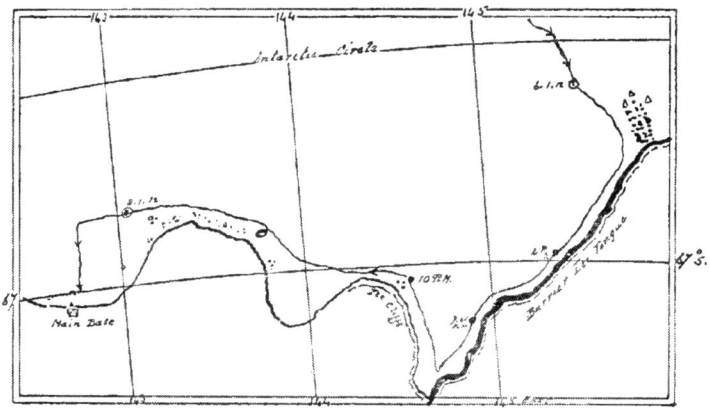

Track of "Aurora" from noon 6.1.12.   Sounding at Noon ₇⁴₅₀
   towards position of Main Base.   ,,   ,, 4 p.m. 398 fathoms on mud.
Noon 8.1.12.   Sounding ₇⁵₀   ,,   ,, 7.45 p.m. ₇⁵₀
   at 1 p.m. course altered to south.   ,,   ,, 10 p.m. 32 fathoms.

8.1.12. At 6 a.m. we proceeded to the westward. The weather was foggy during the forenoon, but it cleared at noon, when the sun came out, enabling us to see several bare rocks toward the head of a bay across which we were steaming. At 1 p.m. the course was altered to approach a large cluster of islets. On coming nearer, a rookery of Adelie penguins was observed on one of the larger islets. Closer in-shore several low islets and grounded bergs could be seen.

3 p.m. It was now beautifully fine with a calm sea and bright sunshine. The ship lay-to and a boat was lowered to take Dr. Mawson and a party to examine the locality. The names of these pioneers—the first to land on the coast of Adelie Land—may be given here: Dr. Mawson, Wild, Madigan, Bickerton, Kennedy, Webb, Bage and Hurley.

The boat returned at 7.30 p.m. Dr. Mawson considered that the locality was suitable for winter quarters, having a sheltered boat-harbour for landing stores, etc., and a considerable area of ground free from ice close to the landing-place. The snow-covered heights to the south and east should furnish a certain amount of shelter from the prevailing winds.

9.30 p.m. The launch was put over, and the whale boat being loaded with odds and ends, both started for the land-

FIRST LANDING ON ADELIE LAND.
Dr. Mawson with pick (January 8th, 1912.)

THE MOTOR-BOAT TOWING TWO WHALE BOATS—LANDING STORES—MAIN BASE—
COMMONWEALTH BAY, JANUARY, 1912

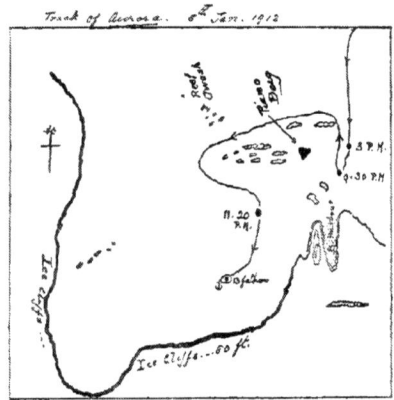

The Anchorage at Commonwealth Bay.
Rough scale 1000 yards to 1 inch.

ing-place. It was arranged that meanwhile the "Aurora" should steam round to the western side of the cluster of islands and proceed to an anchorage which the launch would probably be able to indicate.

When the ship reached the western side of the islands, darkness was setting in and a fresh breeze from the south-east springing up. As the launch was not visible, I began to fear that the engine had broken down, but at 11.20 p.m. both boats were sighted making for the ship. When they came alongside the wind had freshened. In order to save the boats, it was necessary to tow them into shelter under the ice-cliffs. The bottom was uneven, and it was anxious work, but it had to be done. Both boats were taking in water freely.

We dropped anchor in 13 fathoms and gave it 75 fathoms of chain. After dragging a little the anchor brought the vessel up, and preparations were made to relieve the men and to hoist the boats up. Just at this time the launch broke away; the water breaking over the sides affected the motor and caused trouble in starting, but, after drifting half a mile, the launch started by her own power and reached the ship in safety. As the temperature had fallen to 24° F., the pioneers were not sorry to get on board again. Both boats were hoisted and lashed before 1 a.m. (on the 9th inst.).

9.1.12. By the time everything had been made snug lest the anchor should drag, a fresh gale was blowing from the south-east. Occasional squalls of great violence swept down from the ice-slopes. There was no room for the sea to become heavy, or our cable must have parted. The gale lasted all day, the wind lifting sheets of spray under a clear blue sky. The slopes were swept clean, showing the blue glacier ice.

Dr. Mawson had intended to land three separate parties at intervals along the coast line of Antarctica, but, owing to the wind-swept and desolate nature of the newly-discovered land, he decided to attempt only two bases; a main base at our present position, and a second base, under Mr. Wild, at a point not less than 400 miles to the westward. The position of the second base would depend on finding a place to the west of Cote Clarie, where the " Aurora " could approach the coast-line near enough to land the explorers on *terra firma* with the materials and stores necessary to form a wintering station.

10.1.12. The gale lasted until 6 p.m. to-day. Two hours later the weather had moderated sufficiently to allow of the launch making two trips to the shore, whence she returned with blocks of ice, which, on being placed in a large tank, through which a steam coil passes, supplied drinking water sufficient for our present needs.

January 11th and 12th. Landing cargo began in earnest at 8 a.m on the 11th, and continued without interruption from the weather until the evening of the 12th. The wireless masts and hut timber were rafted ashore, and the launch did excellent work during these two days of favourable weather All hands worked vigorously under the subtle influence of the tireless energy shown by our leader. The anchorage is good, and the landing place in the boat-harbour is very convenient. There is quite a large area of bare rock between the water's edge and the foot of the snow slopes surrounding the site chosen for the Hut We appear to have stumbled on a good position for the main base.

January 13th. Dr. Mawson and his party went ashore last night and camped near the boat-harbour, where the launch is quite safe. The force of the wind to-day renders landing from boats impossible, so cargo is being brought on deck to await the return of moderate weather. The wind coming from the slopes causes a choppy sea which does not *affect a ship*, but would soon fill a loaded *boat*. There are a few large bergs aground along the coast, but there is no sign of bay-ice.

This south-easterly gale lasted until noon on January 15th. We were able to resume boat work at 3 p.m., and continue unloading until 9 p.m., when the wind freshened up again,

THE "AURORA" AT ANCHOR, COMMONWEALTH BAY.
Photo. Hurley.

LANDING AT THE HEAD OF THE BOAT HARBOUR.
Photo Hurley.

so that before midnight a strong gale was blowing. When the "Aurora" swings towards the ice-cliffs she has 27 fathoms of water under her stern.

The anchor has held well up to the present, but during one of the violent squalls from the south-east, the cable may part at any moment. I have observed that the wind is generally moderate during the afternoon, and freshens again between 9 and 10 p.m. when blowing from the south-easterly quarter. The only way to land cargo is to take advantage of these lulls and work at high pressure while they last.

January 18th. Early this forenoon the weather fell calm with bright sunshine, after a fresh gale from the south-east all last night. Work proceeded merrily until 10 p.m., when the last load of cargo left for the land. I went ashore with it and took a few snapshots of the locality.

The big bay has been named Commonwealth Bay—a suitable designation for the Main Base of an Australasian Expedition. The site of Hut is on a rocky outcrop known as Cape Denison. The weather this evening was perfect, and the glass steady at 29.10

We have now landed twenty-nine dogs, the air tractor sledge, and twenty-five tons of fuel (at ninety blocks to the ton), and the remainder of the stores. It was a relief to feel that all stores and equipment had been safely landed, notwithstanding the successive gales. The launch has done excellent service, and the convenient boat harbour at Cape Denison has facilitated landing operations.

January 19th. During the forenoon the launch was busy in landing the personal effects of the Main Base Party of eighteen men, including Dr Mawson.

Yesterday afternoon Mr. Webb determined the position of the magnetic station at Cape Denison He reported it as being in latitude 67° 00.5' S., longitude 142° 40' 46" E. He also found the distance from the magnetic station to the anchorage to be 1,037 yards approximately.

The launch made her final trip to the shore, with chronometers and other gear, returning at 3.15 p.m with a load of ice. The cable was hove in to forty-five fathoms, and the launch was hoisted on board.

After heaving short we assembled in the ward room, where Dr. Mawson made a short speech to wish success to those proceeding west to form the second Antarctic party under Mr. F. Wild. A toast to the memory of our predecessors in 1840 was drunk in Madeira which had been part of the sea stock of H.M.S. "Challenger" in 1874. This wine had been presented to the Expedition by Mr. J. Y. Buchanan, one of the scientific staff of that vessel during her famous cruise.

At 8 45 p.m. Dr. Mawson with his staff of seventeen left the ship in their whale boat, amid ringing cheers from all on board the "Aurora"; but not one of us seemed to realize that the time of parting company had come. Our leader had intimated to me that, all being well, he hoped the "Aurora" would return to Commonwealth Bay early in January, 1913, to embark all hands. At the same time, he had every reason to suppose that wireless communication with Australia, via Macquarie Station, would have been established long before that date.

When the whale boat had rounded the entrance to the landing place, the "Aurora" set a course to clear the north-westerly point of Commonwealth Bay.

The last words of the leader were "Good-bye, and *do your best.*" As we on the ship watched the party pulling towards the boat harbour, we began to realize that the real *business* of the Expedition had begun. Difficulty can always be faced, and generally overcome, by patient perseverance, endurance and pluck. But, should men who are striving to "do their best" be overtaken by disaster, their lives have not been spent in vain. They have been the pioneers of progress and noble examples of those who fall "on the field of honour."

CHAPTER VI

A SYNOPSIS OF THE EARLIER VOYAGES ALONG THE COAST LINE OF ANTARCTICA, IN THE AUSTRALIAN QUADRANT

BETWEEN COMMONWEALTH BAY AND GAUSSBERG

"Things gained, are gone,
But great things done—endure"

Swinburne.

BEFORE proceeding with the narrative of our cruise from Commonwealth Bay, a brief summary, compiled from the original reports and charts of those navigators who had preceded us, from 1839 to 1840, will enable readers to follow any reference to these voyages during the cruise of the "Aurora." The "Challenger," in 1873 and the "Gauss" in 1902, had sailed to the southward from Kerguelen Island

From the log of the schooner "Eliza Scott" we have the reported appearance of land, afterwards charted as Sabrina Land.

From the charts and reports of Captain d'Urville we have details of the discovery of Adelie Land in a westerly direction from about longitude 142° E.

From the chart and report of Lieut. Charles Wilkes in the "Vincennes" we obtain a very fair idea of the northern limit of the ice edge (or, as he calls it, "the Icy Barrier") as it existed in 1840. He reported the existence of "High Land" at various places along the coast, but except at Piner's Bay, an "Icy Barrier" and numerous ice islands rendered approach to *terra firma* impracticable.

H.M.S. "Challenger" in 1873 sailed southward from Kerguelen Island, and, crossing the Antarctic Circle, reached latitude 66° 40′ S. in longitude 78° 22′ E.

The chart * indicating the track of the "Gauss" in 1902 shows the discovery of Gaussberg close to the Antarctic Circle in longitude 90° E. This was the western limit of the coastal region to be investigated by the Australasian Expedition.

Very little was known of the coast line between the Main Base at Commonwealth Bay and the territory discovered by the "Gauss" Expedition in 1902, under the leadership of Drygalski.

Changes in the conditions of the coastal ice since 1840 were to be expected. I have tried to show the extent of the change as we approached the Main Base (see page 26). We knew that high land, behind the coastal ice, had been reported at or near the positions shown on the sketch, and that a 60-mile "barrier" had been discovered and charted as Cote Clarie.

The extracts from reports have been given as far as possible in the words of the writers, without comment or criticism. In a later chapter I shall have a few remarks to offer on the voyages of 1840.

(1) *John Balleny in the "Eliza Scott."*

In 1838, Enderby Brothers, a firm of London shipowners, sent out John Balleny in the schooner "Eliza Scott," 154 tons, with the cutter "Sabrina," 54 tons, on a voyage combining sealing with exploration. The ships left Campbell Island, south of New Zealand, on January 17th, 1839, to look for new land to the south. They were stopped by heavy pack in latitude 69° S., longitude 172° E.

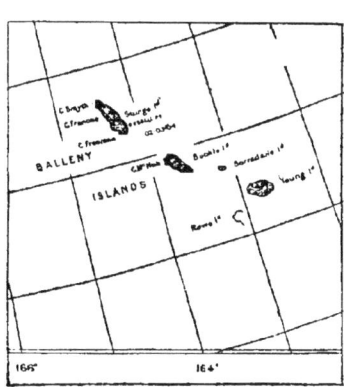

The Balleny Islands.

Proceeding westward they discovered a group of lofty volcanic islands on February 9th, in latitude 66° 44′ S., longitude 163° 11′ E. Balleny followed a westerly course between the parallels of 63° and 65° S., keeping a lookout for new land. During a gale the ships separated, and the "Sabrina" was lost with all hands. Balleny reached England safely and reported an appearance of

* See page 41

land in latitude 65° S. longitude 122° 44′ E. This was claimed as a discovery and charted as " Sabrina Land," after the unfortunate cutter.

During the year 1840, three national naval expeditions explored certain portions of the coastal region close to, or just south of, the Antarctic Circle. The special object of each expedition was to locate the South Magnetic Pole, the approximate position of which (from the theoretical investigations of Gauss), was supposed to be in latitude 66° S., longitude 146° E.

(2) *The French Expedition.*

Two corvettes under the command of Captain Dumont d'Urville sailed from Hobart Town on January 1st, 1840. The " Astrolabe " and the " Zelée " (Captain Jacquinot) were re-fitting at Hobart Town in December, 1839, after a long voyage in the Pacific. Captain d'Urville heard that an American squadron had reached Sydney, and was then preparing to cruise in search of the South Magnetic Pole. He also heard that a British expedition was being fitted out, and that the vessels, under the command of Captain James Clark Ross, were expected to reach Australian waters during the coming year. Dumont d'Urville, therefore, decided to make a dash for the Pole for the honour of France!

Sailing south, he sighted land on the evening of January 20th. The following day the weather was fine, and good observations gave the position of the " Astrolabe " at noon as latitude 66° 30′ S., longitude 140° 41′ E. (of Greenwich)—about 6 miles from snow-covered land sloping to a height of about 1,500 ft. Many large bergs lay off the coastal ice-cliffs. Observers were landed on a large berg, and three hours were spent in taking magnetic observations.

In the afternoon, a rocky islet, lying close to the foot of the ice-cliffs, was seen from the ship. Two boats pulled in-shore and returned laden with geological specimens, a piece of dry seaweed, and several penguins. Some smaller rocky islets were seen round a projecting point named " Pointe Géologie." The islet on which the landing was made was called " Rocher de la Découverte."

This new land was called " Adélie Land," after Madame d'Urville.

The coast was followed to the westward as far as longitude

138° E., where dense pack ice was met extending from the coast to the N.E. The pack ice was followed until January 23rd, when a violent gale drove the vessels to the north.

On January 29th the ships had worked back to latitude 64° 45′ S., longitude 135° 50′ E. (approximately), where the pack was met running nearly east and west, fringed by ice islands and bergs. Some of the sailors thought that land was visible to the south over the ice, but the navigator writes in his report : " I am almost certain that ' Adélie Land,' of which we traced about 150 miles, extends to our longitude, but *too far to the south* to be visible from our position to-day."

The ships then followed a course a little to the north of west.

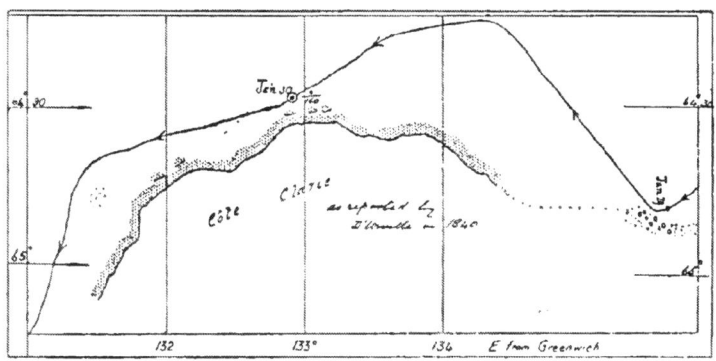

Côte Clarie from Durmont D'Urville's chart.

On January 30th the corvettes were within 4 miles of an ice-cliff with vertical front and flat top which rose from 120 to 130 feet above the water. Here and there large bergs lay in front of, or along the foot of this ice wall, but the sea was nearly clear of ice in the open.

A sounding at noon gave no bottom at 160 fathoms. There was no appearance of high land to the south of this barrier, which was followed for about 60 miles. D'Urville formed the opinion that this huge mass rested on a base of land, or rock—perhaps on scattered shoals—at some distance from the mainland.

It was called " Côte Clarie," after Madame Jacquinot.

After following the barrier towards W.S.W. for twelve hours, the direction changed to S.W. This course was followed during the hours of darkness, and at daylight on

January 31st nothing could be seen of the ice-cliff. A chain of ice islands had taken its place. Later, heavy pack ice was seen from aloft extending far to the W. and N.W.

February 1st. The ships set a course for Hobart Town which was made safely on February 17th. The last ice was passed in latitude 57° S.

(3) *The United States Expedition.*

Two sloops and a gun brig of the American Squadron, under Lieut. Charles Wilkes, sailed from Port Jackson on December 26th, 1839.

The " Vincennes " (780 tons), the " Peacock " (650 tons), and the " Porpoise " (230 tons) sailed under instructions " to attain as high a southern latitude as possible between 160° and 45° E. proceeding from east to west."

Lieut. Wilkes issued instructions to the " Peacock," and the " Porpoise " to rendezvous along the icy coast-line, in case of separation from the flagship.

On January 19th the " Peacock " was nearly lost in the ice. After being patched up, she returned to Sydney for repairs. The information obtained by the " Porpoise " on this western cruise added nothing material to that obtained by the " Vincennes."

Extracts from Wilkes's Report.

January 25th, 1840. Disappointment Bay, about 25 miles wide. We explored it to a depth of about 15 miles. The whole bay was enclosed by a barrier of ice from N.W. to N.E. Land appeared both to east and west, 67° S. and 147° 30′ E. is the position of this " Bay."

(See sketch map on page 26.)

January 28th. Land was seen distinctly in longitude 141° E., but the " Vincennes " was driven north through a berg-laden sea by a strong south-westerly gale.

On the 30th the ship had worked back to 66° 45′ S^{m} 140° 02′ E. where a sounding on hard bottom gave a depth of 30 fathoms, about 1½ miles from the coast-line. The weather was too rough to get the boats out, and Wilkes had to abandon any attempt to land on bare rocks, but he named the place " Piner's Bay," being unaware that the French ships had visited the locality *nine days* previously.

The " Vincennes " held on to the westward from February 3rd to February 6th, encountering strong gales with snow.

February 7th. The ship was following an icy barrier 150 ft. high, while, beyond it, outlines of high land could be dis-

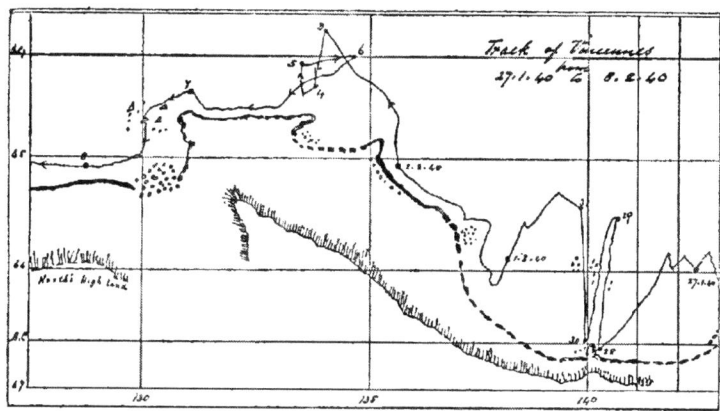

Track of the "Vincennes" from the Ross tracing.

tinguished. At 6 p.m. the barrier was trending to the south and the sea was covered with bergs.

"I now hauled off till daylight to ascertain the trending of the land more exactly. I place this point which I have named 'Cape Carr,' after the First-Lieutenant of the 'Vincennes,' in latitude 64° 49′ S., longitude 131° 40′ E."

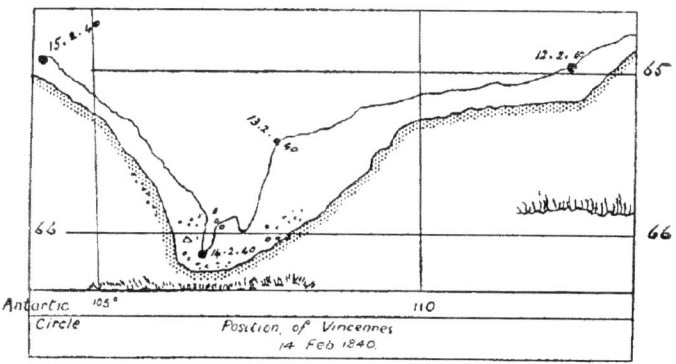

Track of the "Vincennes" off Knox Land.

EARLIER VOYAGES ALONG COAST LINE

On February 12th, land was sighted trending S.S.E. to S.W. from 18 to 20 miles distant. Sounded in 250 fathoms, *no bottom*. "I put this land in about 65° 20′ S. and 112° 16′ E. (" *Budd's High Land*.")"

February 14th. "Beating in towards land until 11 a.m. Stopped by ice when 7 to 8 miles distant. Day very clear and land very distinct—snow-covered, 3,000 ft. high, extending 75 miles. Landing made on floe."

The land is marked on the official chart of the Expedition as " *Knox Land*."

February 15th. "The sea has been smooth for the last few days. No swell. I began to think we might find large bodies of ice to the north. Cook was stopped in 1773 200 miles north of our position."

February 16th. "The barrier of ice trended to the north. We passed through a number of ice-islands stained with earth."

February 17th. "To-day we observed the barrier extending in a line ahead, running north and south as far as the eye could reach. Appearances of land were also seen to the S.W. and its trending seemed to be to the northward. We are thus cut off from any further progress to the westward. We were now in latitude 64° 01′ S., longitude 97° 37′ E."

Repulse Bay, position of "Vincennes" 17·2·'40.

(The appearance of land referred to is named " *Termination Land* " on the official chart.)

February 20th. "Sounded in 850 fathoms. Wind west. We are now nearly in the latitude where Cook was stopped in 1773."

On February 21st the barrier was seen to be trending westward : " Indications of approaching bad weather and the proximity of so many ice islands made me decide to turn the head of the vessel northward."

The "Vincennes" reached Sydney on March 11th. Wilkes refers to his original chart as follows :—

"The tracing of the icy barrier attached to the Antarctic Continent discovered by the U S Exploring Expedition 1840, was communicated by Lieut. Wilkes to Captain James Ross. As we sailed along the 'Icy Barrier,' I prepared a chart, laying down the land, not only where we had actually determined it to exist, but also those places in which every appearance denoted its existence. . . . I had a tracing made of this chart, which was forwarded to Captain Ross through Sir G. Gipps at Sydney."

(4) *The British Expedition.*

Two naval vessels of great strength, the "Erebus" and the "Terror," reached Hobart Town in August, 1840, under the command of Captain J. C. Ross, R.N., who had reached the North Magnetic Pole in 1831.

Captain Ross heard of the discoveries of d'Urville and Wilkes, and was annoyed that others had chosen the route marked out for the British Expedition. He decided "not to follow in the footsteps of the Expedition of any other nation," and resolved to penetrate south on the meridian of 170° E., where Balleny had found open sea in latitude 69° S., but he did not forget to acknowledge the courtesy of Lieut. Wilkes in "sending a letter accompanied by a tracing of his original chart."*

This tracing was published as an appendix to *Vol. I. Ross*. It is on a larger scale than the official chart. No names are marked, but the track of the "Vincennes" is clearly shown.†

(5) *The Challenger Expedition*—1872–74.

H.M.S. "Challenger," 2,306 tons, a wooden ship with auxiliary steam power, was the first steamer to cross the Antarctic Circle, on her voyage from Cape Town to Melbourne via Kerguelen, in December, 1873 She sailed south-

* See p 115, Vol I, *Ross*

† Sketch maps in this volume showing the track of the "Vincennes" are from this tracing

ward from that island, reaching latitude 66° 40′ S., longitude 78° 22′ E.

(6) *The German Antarctic Expedition of* 1902.

The track from Kerguelen Island to Winter Quarters is taken from a map published by the Geographical Society to illustrate a paper by Drygalski who describes the discovery of Kaiser Wilhelm II Land.* "During the night on Feb-

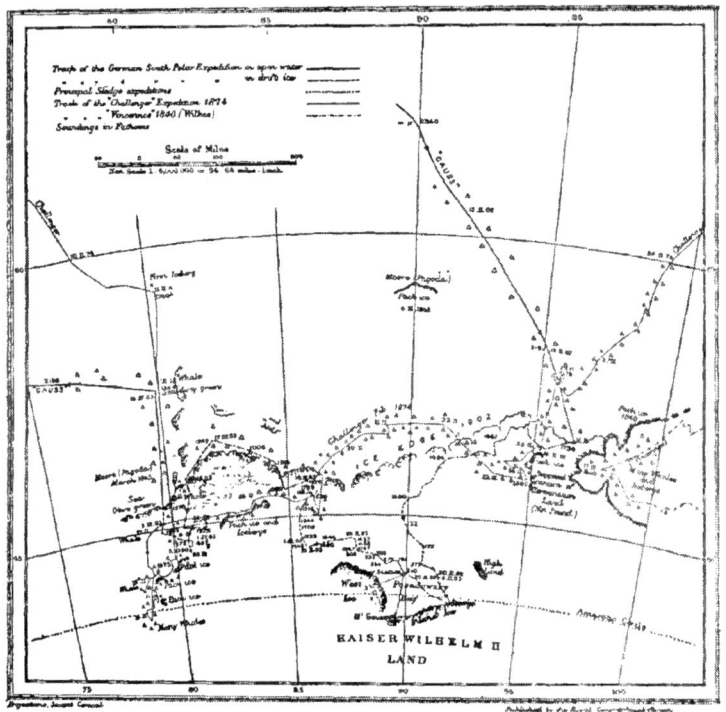

Track of the "Gauss."

ruary 21st the weather improved, and when the short night yielded to day there lay the land—a new land—before our eyes.

Whatever feelings may have agitated us at this sight— whether grief that a stop was so soon put to our ambition to advance farther south, or joy at the actual result, the

* The German Antarctic Expedition, by Dr. Eric von Drygalski.—*The Geographical Journal*, August, 1904.

happy solution of a great geographical question—one thing is certain, that for the moment all other feelings gave way before the imposing view of land. There it lay in its quiet solitary grandeur, never before beheld, never before set foot on. All was ice-clad, still, that it was land was shown beyond all doubt by the very forms affected by the ice. For, looking towards the coast, we could see the uniform surfaces which sloped down from the south in broad, smooth undulations, branching off and developing glaciers, such as are conditioned by the forms of a firm substratum. The coast itself was a high vertical wall of ice, too steep to be approached in whatever direction we turned our eyes, only somewhat diversified high up and with an approximate east-to-west trend—in fact, such another ice-cliff as in his time confronted Ross on the southern margin of his Ross Sea.

A landing on this icy barrier was out of the question. Hence we resumed such operations as, in the absence of ice-free tracts, might lead to some conclusions regarding the substratum of the ice-cap—that is the character of the land itself. We accordingly took soundings, fished with the drag-net, and made magnetic observations. Then we continued our course in the direction of the west, since from the spot where we had struck land, and from what we had already seen of its nature, we had drawn a sufficiently accurate inference regarding the section of Wilkes Land which lay east of us "

The "Gauss" was beset on Febuary 22nd, 1902. About the middle of March, when the position of the vessel was not likely to change, a sledging party travelled south and reached land in $3\frac{1}{2}$ days. The party saw a black hill (volcanic cone), rising before them to a height of 1,200 feet amid the surrounding ice and snow. This hill was named "Gaussberg," its position from Winter Quarters being estimated at 50 miles. On March 29th Drygalski went up in a captive balloon and remained aloft at a height of about 1,500 feet for two hours; from this elevation land was observed to the eastward. Ten years later a party from the Western Base of the Australasian Expedition proved this land to be an island, which was named after the discoverer Drygalski Island.*

* See page 104

CHAPTER VII

THE FIRST ANTARCTIC VOYAGE (*continued*)

COMMONWEALTH BAY TO THE WESTERN BASE AND BACK TO HOBART

WE had left our leader and his comrades to settle down for the southern winter, in their windy quarters at Cape Denison. It was the business of the " Aurora " to discover a suitable locality for the second Antarctic Base.

About midnight on January 19th, we reached the western point of Commonwealth Bay, where a long chain of grounded bergs stretched away to the north-east The coast line was about five miles distant, and many small islands close in-shore showed up as black patches against the ice cliffs which appeared to be about 100 feet high. Above these, the snow slopes rose to a height of 1,300 feet. The weather was fine and clear, enabling us to see for a long distance. We found that owing to the proximity of the magnetic pole, our compass was comparatively useless for steering courses.

20.1.12. At 4 a.m. we sighted a long reef extending from the coast-line in a northerly direction. At reduced speed, we felt our way round this obstacle; soundings, at frequent intervals, ranged from 32 to 100 fathoms on rock and sand bottoms. At 8 a.m. we resumed a westerly course, steering for the next point of the land. Fortunately, the weather was clear when this reef was first sighted, but the incident made us all realize that our business was to *make* a chart of the coast-line, and not to depend on one, as in better-known latitudes. As we advanced from point to point westward, at about 12 miles from the ice-cliffs, we sighted large bergs, some hummocked and crevassed, others tabular, like those in the Ross Sea, but nearly all were portions of the ice-cap which had been pushed out and then broken off from the shore-ice.

Our position at noon was in latitude 66° 33′ S., longitude 140° 28′ E. A sounding was taken in 308 fathoms on mud.

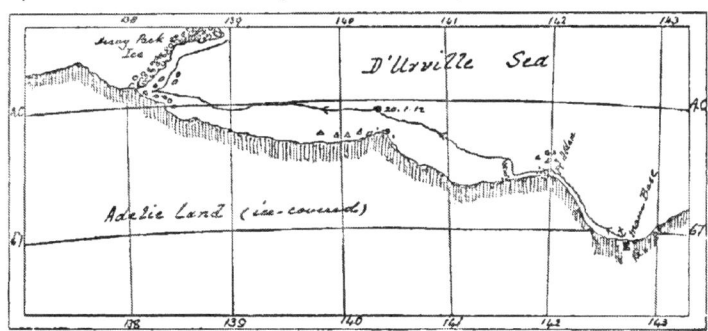

Track of "Aurora" off Adelie Land.

Captain D. d'Urville (in the "Astrolabe") records his position at noon on January 21st, 1840, as being in latitude 66° 30′ S., longitude 140° 41′ E. (of Greenwich). A landing was made on a rocky islet close to Pointe Géologie. The French Navigator was favoured by fine weather for twenty-four hours after sighting Cape Discovery, and his description is remarkably accurate.

Lieut. C. Wilkes (in the "Vincennes") reached an indentation in the coast line, which he called Piner's Bay, in latitude 66° 45′ S., longitude 140° 2′ E. on January 30th, 1840. The ship approached the coast-line to within 1½ miles, and a sounding was taken in 30 fathoms on hard bottom, but the advent of a gale prevented a landing being carried out, and the "Vincennes" was driven to the North.

At 10.30 p.m. we found our way westward barred by a line of bergs, close to the shore, and by a field of berg-laden pack-ice extending to the north-east. The coast-line of Adelie Land could be traced for some distance in a westerly direction, and, where it disappeared from view, there was an appearance of a barrier running in a northerly direction. As we followed the pack-edge on January 21st and 22nd, the appearance of a barrier beyond the pack was observed from aloft. The existence of such a barrier possibly resting on a line of reef, similar to the one met with near Cape Discovery, would account for this ice-field occupying practically the same position in 1912 as it did when seen by d'Urville in 1840. We were unable to see any trace of the "high land" reported by Lieut. Wilkes, as lying to the west and south-west, beyond the ice.

The course followed until the morning of January 23rd is shown on the sketch.

THE FIRST ANTARCTIC VOYAGE (continued)

22.1.12. There has been a heavy swell to-day from the north and north-east. Very little change has taken place in the general position of this field of pack since first reported.

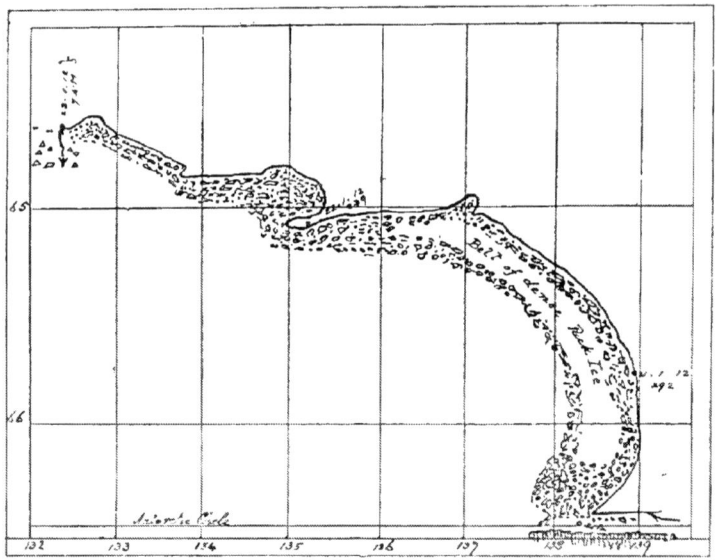

Track of the "Aurora" to 23.1.12

There was a gorgeous sunset this evening. The weather was quite calm, and a bright ice-blink rose to about four degrees above the horizon. All around were glistening bergs, displaying the most exquisite colours. A solemn silence over all was broken, at intervals, by the hoarse cry of a penguin.

View of Ice-cliffs, Côte Clarie, from chart of D'Urville, 1840.

23.1.12. 1.30 a.m. We cleared the pack and steered due west until 7 a.m., expecting to sight Cote Clarie about four miles to the south of our course. On making longitude 132° 30′ E. we stood south, and shortly afterwards passed over the charted position of Cote Clarie. The water here was clear of pack-ice but studded with huge bergs. The great barrier which had been followed by the French and American ships, for a distance of about sixty miles, had vanished. The weather was remarkably fine and clear, but nothing was visible to mark its former position except a collection of immense bergs.

<small>Captain D d'Urville writes · "All day long on January 30th (1840), the ships sailed along a vertical cliff about 120 feet high, quite flat on the top with no signs of higher land beyond (i e , to the south)"</small>

At 10 a.m., having passed to the south of the charted position of Cote Clarie, the course was altered to S. 10° E. true. Good observations placed us at noon in latitude 65° 2′ S. longitude 132° 26′ E., with a sounding of 160 fathoms on mud and small stones. We sailed over the charted position of *land* east of Wilkes's "Cape Carr," but no trace of land was seen in the locality. At 5 30 p.m. land was sighted to the southward ; snow-covered hills similar to those already seen in Adelie Land, but of greater elevation. We continued on the same course towards this land until 8.30 p.m., when we were stopped by heavy floe-ice extending right up to the coast. We estimated the distance of the nearest point on the line of coast at 20 miles. A sounding gave 230 fathoms on sand and small stones. Some open water could be seen to the south-east; so we tried to approach the land in that direction.

24.1.12. At 3 a m. we had reached a point within 12 miles of the coast, when further progress was stopped by the ice. We could see that the slopes were heavily crevassed and they *appeared* to run down into the sea instead of ending in a barrier-face, but, as there were many bergs close to the shore, and as the ship was at a considerable distance from it, the precise nature of the coast-line could not be determined.

We turned back on a north-westerly course, and our noon position was slightly to the north of our position at 5.30 p.m. on the 23rd. We sounded in 170 fathoms on mud. We were surrounded by every kind of berg. Some were from 180 to 200 feet high. During the afternoon we steered west until 4 p.m., when the course was altered to south-west, as I hoped to get in with the land which was visible on the southern horizon. At 8 p.m. the sky was very clear to the southward, and the land could be traced for a considerable distance until it disappeared in south-west true.

STEAMING THROUGH LOOSE PACK.

THE WAKE OF THE VESSEL THROUGH LOOSE PACK.

THE FIRST ANTARCTIC VOYAGE (*continued*)

The pack edge now trended to the north, and we were compelled to follow that course. Heavy cumulus clouds were approaching from that quarter; they were gradually blotting out the clear sky to the southward.

During the days which followed violent gales and heavy seas prevailed, and we were blown some distance to the northward. Owing to thick weather, our position was difficult to fix, even approximately. During this period we had some narrow escapes. The wind was at times so violent that, in a heavy sea, the ship would not answer her helm. Under such conditions the loom of a long berg showing up on the lee bow makes one feel uncomfortable. However, I must say that the "Aurora" behaved admirably, as she invariably does in heavy weather.

We picked up the edge of the main coastal pack again on the 29th, in latitude 64° 56′ S., longitude 126° E. (approximately),

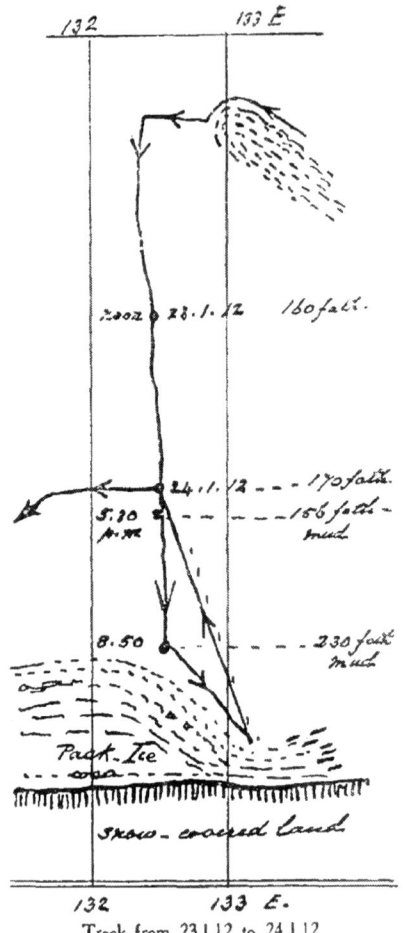

Track from 23.1.12 to 24.1.12

but good observations for position were not obtainable owing to fog, until January 31st, when our position (at noon) was in latitude 66° S., longitude 119° 30′ E., with a depth of 340 fathoms on mud bottom.

31.1.12. The appearance of land (reported by Balleny

in 1839 as being close to the Antarctic Circle in longitude 123° E.), afterwards charted as "Sabrina Land," was not to be seen. We sailed over the position indicated, but there was no appearance of land in the vicinity.

At noon we should have been about ten miles north of "Totten's high land," as reported by Wilkes. Although the weather was clear, nothing could be seen over the heavy floe-ice to the southward except a faint blue line on the southern horizon. This may have been a lead of water, or it may have been land, but in any case we could not approach it owing to the ice.

We made fast to a large floe and secured a good supply of ice before resuming our course along the edge of the pack.

1.2.12. Our position at noon was in latitude 65° S., longitude 116° E., with a depth of 927 fathoms on mud and stones. There was a noticeable absence of ice-islands within the limited range of vision. A marked improvement in the compass is apparent, for which I feel truly thankful. We are now getting away from the area of low, directive, magnetic force.

3.2.12. We were on this day in the longitude of Knox Land, where a landing had been effected in 1840 on a large floe, about eight miles from the coast-line. Lieut. Wilkes reported that a nearer approach had been barred by the ice, and that snow-covered land, rising to a height of 3,000 ft., was seen running from east to west for a distance of 75 miles. I hoped that a landing place would be found in this locality for Mr. Wild's party. Time was running on towards the end of the season for settling in winter quarters, so I decided to make a determined effort to push through the belt of pack in the direction of Knox Land.

Our position at noon was 65° 39′ S. and 108° 35′ E., with a sounding of 300 fathoms on sand. The pack being fairly loose on the edge we pushed into it on a southerly course. Soon afterwards progress was stopped by floes of great thickness. There was no appearance of open water to the south I therefore decided to retreat and try to penetrate farther west.

4.2.12. A second effort to push south was made to-day in longitude 106° 52′ E., but after advancing a few miles we were brought up by heavy floes, many of which showed signs of severe pressure. This effort was made in latitude 65° 7′ S., where Wilkes had found open water 50 miles further south.

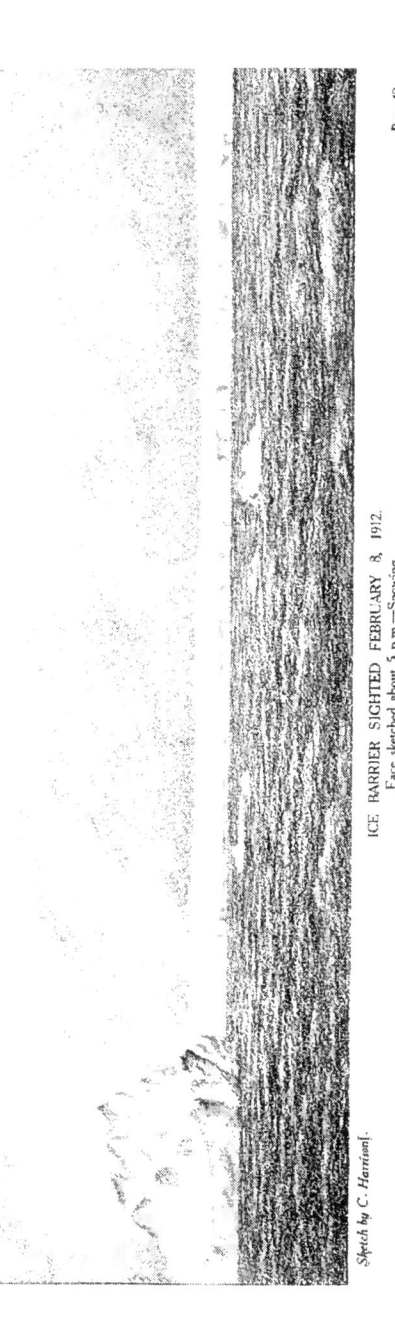

ICE BARRIER SIGHTED FEBRUARY 8, 1912.
Face sketched about 5 p.m—Snowing

Sketch by C. Harrison].

THE FIRST ANTARCTIC VOYAGE (*continued*)

After a third attempt farther west, during which the "Aurora" received some hard knocks, and was only extricated with great difficulty, I came to the conclusion that it would be hopeless to try and work through over fifty miles of such heavy ice. We therefore continued to follow the pack-edge westward towards the position assigned by Wilkes to "Termination Land." The course made was approximately north-west.

8.2.12. At 7.40 a.m. a glacier-tongue was observed extending in a north-westerly direction. It was about eighty feet high and sloped back from the top of the ice-cliffs like the roof of a house. Following this, we found ourselves at noon in latitude 64° S., longitude 98° E. (approximately).

We reached the north-western end of the tongue about 8 p.m., where a sounding gave 850 fathoms and *no bottom*. The tongue then trended away sharply to the S.S.E. After passing through some loose ice, we

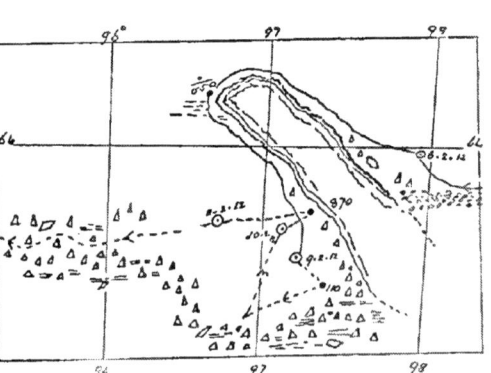

Sketch of course followed by the "Aurora."

followed the western side of the tongue to the southward.

9.2.12. The ship is still following the tongue. At 5.30 p.m. the water had shallowed to 110 fathoms on a bottom of mud and small stones. A line of grounded bergs and heavy floe-ice was observed ahead extending from the edge of the ice-cliffs round to north-west true. The barometer was falling rapidly, so I decided to remain under the lee of the tongue until the weather improved. With all these bergs and floes round our position, it would be senseless to leave shelter to try experiments.

10.2.12. A blizzard raged all day and we were fortunate in having the barrier to windward. At noon we got a depth of 102 fathoms on coarse sand. When about twelve miles to the eastward of our position at noon (close to the ice-cliff), we found a depth of 870 fathoms on sand.

11.2.12. The weather having moderated a little, we made an attempt to steer south, but the weather became thick, and floes were observed ahead, so we turned back to north-east at slow speed. In the afternoon we steered to the west, through more open water.

12.2.12. At 4 a.m we stood to the south-west. At noon the sea was thickly studded with bergs and large floes. There was some open water between them through which we managed to work our way. Good sights for longitude were obtained. At 5.30 p m. we sounded in 235 fathoms on mud, in latitude 65° 6′ S., longitude 94° 2′ E. We kept on through open water in a southerly direction until midnight.

13.2.12. At 1 a.m. we were brought up by a detached mass of floe As it was rather dark, the ship was put head-on to the ice to await daylight. After steaming round the floe, we stood south in absolutely open water. At noon the depth had increased to 500 fathoms, in latitude 65° 54¼′ S., longitude 94° 25′ E. Shortly after noon, an ice-tongue was observed on the port bow; the height being about 100 feet. This was followed to the southward. Early in the afternoon, high snow-covered land was seen trending nearly east and west.

At 4.30 p.m. we reached the edge of old and heavy fast-ice, which appeared to extend for some twenty miles right up to the land. At this point, a sounding gave 250 fathoms on mud.

Mr. Wild and a comrade landed on the ice and made a short journey towards the new land. He reported that the coast was fully twenty-five miles from the point they had reached, and that the ice was evidently breaking up. As it was out of the question to disembark a party on ice of this description, we anchored to the floe. A number of Emperor penguins and Weddell seals were noticed on the floe, during the afternoon.

14.2.12. There was nothing to be done except to return northward, so we kept under the tongue, steaming slowly and searching for a landing-place. We found a low place where the ship could have come alongside with a depth

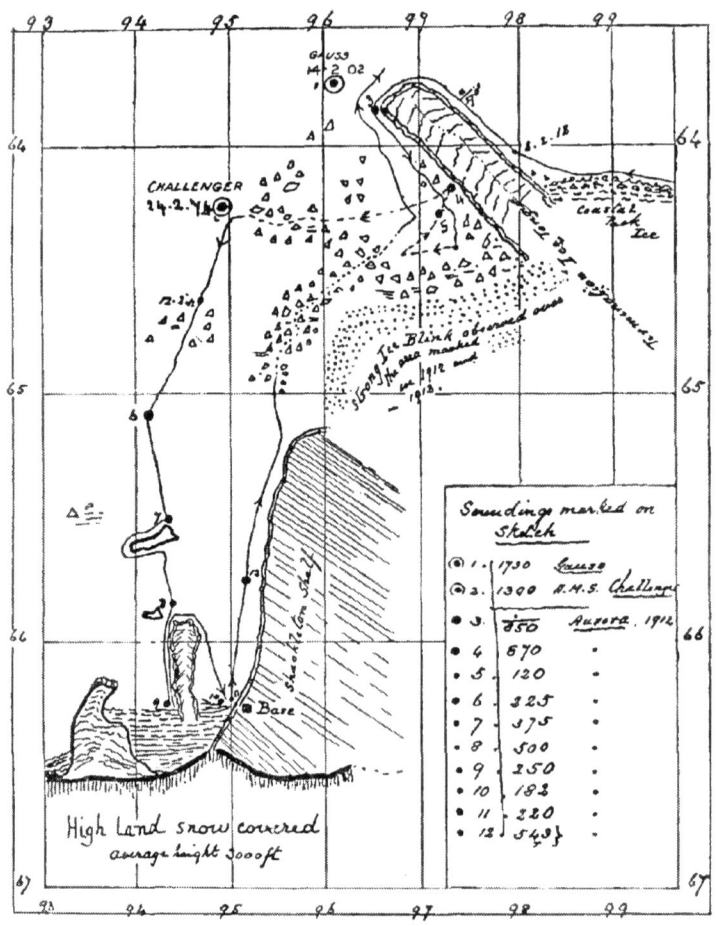

Landing of Wild's Party, 1912. Track of "Aurora" from 8.2.12. to 23.2.12.

of 160 fathoms. A little later we found two channels leading right into the tongue. As this seemed to indicate that the mass of ice was practically adrift, it would not serve as a landing-place. As we steered along the face we saw large pieces breaking away and drifting to leeward.

15.2.12. At 4 a.m. we cleared the northern end of the tongue and stood to the eastward. This mass of ice was about 24 miles long, but narrow in proportion to its length. There was a strong swell from the north-east, and the temperature had fallen considerably.

At 8.45 a.m. a high, barrier formation was observed from aloft, trending about north and south. It met the floe ice in a south-easterly direction, from the position of ship at the time, so the course was altered to south-east.

At noon we were up to the point where the ice-wall met the floe-ice. We sounded in 182 fathoms. Mr. Wild and party landed on the floe and were able to get to the top of the overhanging cliff. The party soon returned, and Wild reported that at this place the barrier consisted of blue glacier-ice; he described it as being like the Great Beardmore Glacier. He considered that excellent winter quarters could be erected about 600 yards from the top of the ice-cliff, which rose to a height of sixty feet. This spot was about seventeen miles from the nearest point of the coast-line; and a flat surface covered with snow stretched away to the east and north as far as the eye could see.

The "Aurora" was then brought close up to the sea-ice at the foot of the cliff, and the work of landing the stores and hauling them to the top of the cliff, by the help of a "flying fox," began at 4 p.m.

The mean of a number of sights gave the position of the ship as in latitude 66° 18′ 28″ S., longitude 94° 58′ E.

On February 19th the landing of stores and equipment was completed. Twelve tons of coal was the last item sent up to the top of the cliff.

20.2.12. To-day the weather is fine—a great improvement on the keen wind and drifting snow which we have had for the last two days. The ship is getting very light, and we must not linger here, as our stock of coal is reduced to ninety tons, and we have a voyage before us of 2,300 miles to Hobart.

THE "AURORA" OFF WILD'S BASE

THE FLYING-FOX USED FOR HAULING STORES FROM THE FLOE-ICE TO THE TOP OF CLIFF

WILD'S BASE ON THE SHACKLETON SHELF. [Photo, Gillies.

LANDING STORES AT WILD'S BASE.
Showing the top of the Cliff. [Photo, Gillies.

We did not get the coal ashore a moment too soon, as, this morning, the sea-ice traversed by our sledge-track broke away and drifted off in a north-westerly direction. This afternoon it is drifting back under the influence of a tide or current.

I went up on the glacier with Wild this afternoon. It is somewhat crevassed for about 100 yards inland, and thence a flat surface of great extent stretches away to the eastern horizon.

21.2.12. At 7 a.m. we said farewell to Wild and his seven comrades, who scrambled over the rail with their blankets on their backs They made their way across the sea-ice towards the heap of stores upon the glacier, which represented all they had to depend upon for the next twelve months—a black " oasis " in a white waste of snow ! Was this an ice-shelf, attached to the land, on which we were leaving them ? or would it, and they, have " gone to sea " before the arrival of the " Aurora " next year ?

We steamed along the ice-cliff until 8 p.m., when we were in latitude 65° S. Here the barrier trended off to the E.S.E. In character and appearance, the formation was similar to the Ross Barrier, except that it was not quite so high. From the site of the Western Base northwards the height ranged from 60 to 70 feet.*

We are now leaving the zone of bright, clear weather and entering the region of fog and gloom. There appears to be a well-defined line just north of the coast-line, where the sky is always overcast; I noticed a similar phenomenon when off Adelie Land.

11 p.m. We are now passing a line of grounded bergs, and I see some heavy floe-ice. Fortunately it is calm, but being dark, it is difficult to make out a passage through the bergs and floes

22.2.12. I cannot explain how we managed to clear the obstructing ice between 11 p.m. last night, and 3 a.m. this morning. We tried lying-to, but it soon became evident that some of the bergs were moving and would probably hem us in, so we pushed this way and that way, endeavouring, at any cost, to retain freedom of action. About midnight

* At a later date, this immense ice-formation was found to cover an area of several thousands of square miles, and was named the " Shackleton Ice-Shelf "

we were trying to clear the loom of a large berg on the starboard bow, when a wall of ice appeared to rise out of the haze, stretching right across our course. There was no room to turn, so "full speed astern" was the only alternative. The signal was answered *immediately*, or we must have crashed into this mass of ice. Until daylight, it was "ice ahead," "ice to port," "ice to starboard," in fact "ice everywhere." The absence of wind saved us from disaster, and we had a providential escape. It was a great relief when daylight appeared showing clearer water to the northward.

We made Termination Ice-Tongue this afternoon and reached its north-western point about 8 p.m. As the wind was freshening, with a falling barometer, I decided to shelter under the Tongue until daylight.

23.2.12. This morning we met a few bergs off the end of the Tongue as we steered nearly due north in a fresh breeze with a high sea. No pack-ice was visible.

We discovered that one ballast tank will not hold water, and that the other is leaking. Meanwhile, we require every pound of ballast that can be had, because an empty ship in these latitudes is no joke. The ashes from the stokehold are being wetted and put down below again, while the Chief Engineer is doing what is possible to repair the tanks.

The last ice was seen in 59° 30′ S. on February 29th. During the next three days we had a succession of heavy gales, but the Chief Engineer had succeeded in repairing the ballast tanks, so we were able to carry twenty-two tons extra, which made a great difference. On March 5th we got the launch down on the main deck from off the top of one of the deck houses, and as the tanks continued to hold water we had many things for which to be thankful. The "Aurora" is doing well and has beaten the attempts of the weather to turn her over during several heavy gales.

Hobart was reached safely on March 12th. On our way up the Derwent, we passed the famous Polar ship "Fram" as she lay at anchor in Sandy Bay. (The pilot had told us of Amundsen's success.) Flags were dipped, and a hearty cheer was given for Captain Amundsen and his gallant comrades who had raised the siege of the South Pole. The "Fram" is a three-masted schooner, 117 ft. long, built in 1892.

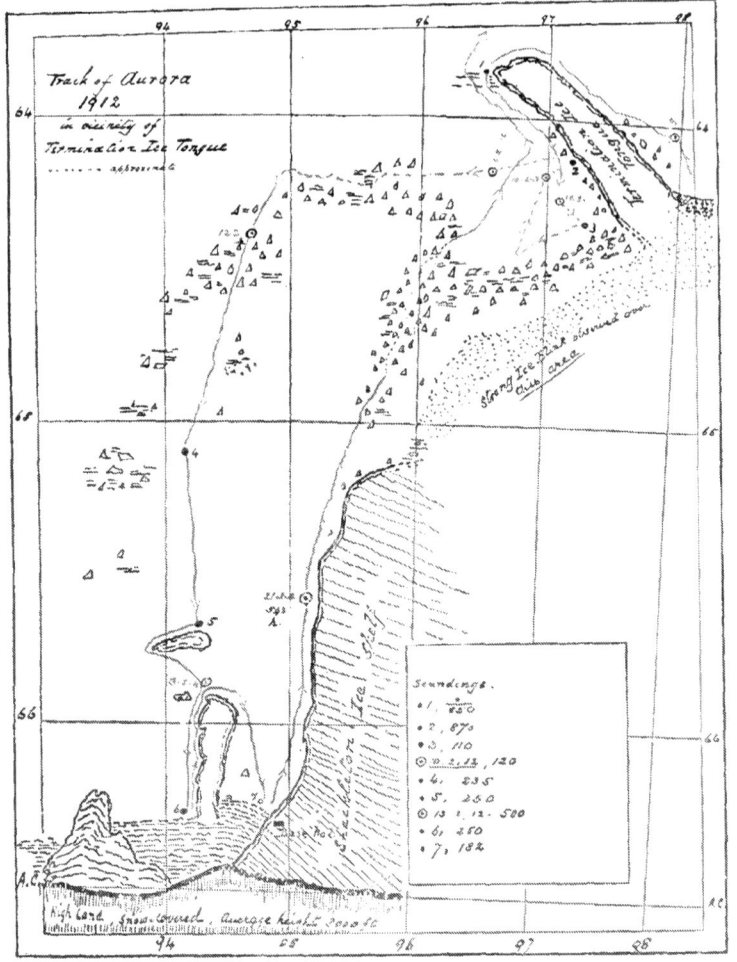

Track of the "Aurora," 1912.

THE "FRAM" AT HOBART, TASMANIA.

On the following day I had the honour of conveying the warm congratulations of the Australasian Antarctic Expedition to the Norwegian leader and his party on their very remarkable feat. The Expedition had been organized with the greatest care by an explorer of long experience. His efforts had been seconded by devoted comrades, and Fortune had smiled on their united efforts

When we made fast to the pier at Hobart, only nine tons of coal were left. The draught of the ship was 12 feet 11 inches forward, 15 feet 9 inches aft.

We were warmly welcomed at Hobart, and enjoyed the generous hospitality of the citizens. Our stay in port was not a long one, as the "Aurora" had to leave for Sydney to be docked and refitted.

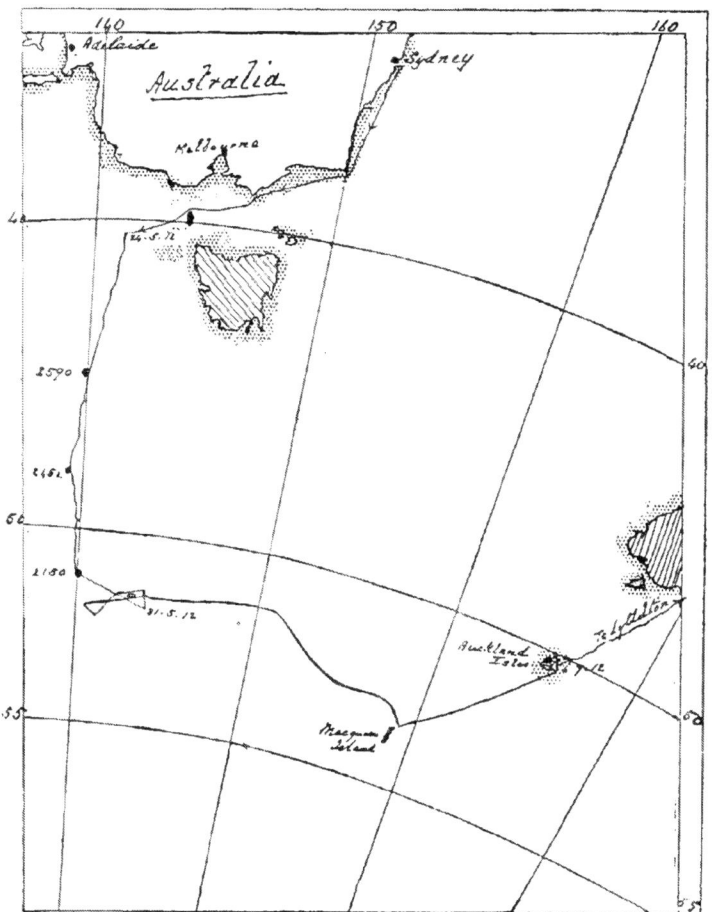

Track of "Aurora"—Sydney to Lyttelton—Winter Cruise.

CHAPTER VIII

A WINTER CRUISE

SYDNEY TO PORT LYTTELTON VIA BASS STRAIT, MACQUARIE ISLAND, AND THE AUCKLAND GROUP

"Where the besom of God is the Wild West Wind
That sweeps the sea floor white"

THE "Aurora" was docked and overhauled in the Government dockyard at Sydney. A new steam windlass was fitted, and the special equipment required for deep-sea work was taken on board. These preparations were completed by May 17th, when we sailed from Port Jackson to coal at Port Kembla. We were ready for sea on May 20th and commenced our winter cruise.

Mr Waite, Curator of the Canterbury Museum, accompanied us as biologist. Mr. Primmer acted as cinematographer. He hoped to secure some moving pictures of popular interest, but the light even at midday in southern seas at this season of the year proved to be most unfavourable.

My instructions were to sail over the position assigned to the Royal Company Islands,—on a French chart,—before steering for Macquarie Island, so I decided to follow a westerly course through Bass Strait as far as longitude 141° E ; this should place the ship well to windward of the charted position in latitude 52° S , longitude 143° E.

Before narrating our experience during the voyage, I shall briefly describe some of our deep-sea appliances.

A glance at the deck plan of the "Aurora" will show the positions of the various apparatus employed for sounding and trawling. Brief descriptions are added on the construction and method of working of these appliances.

Sounding Apparatus.

(1) A Lucas sounding machine (for depths up to 6,000

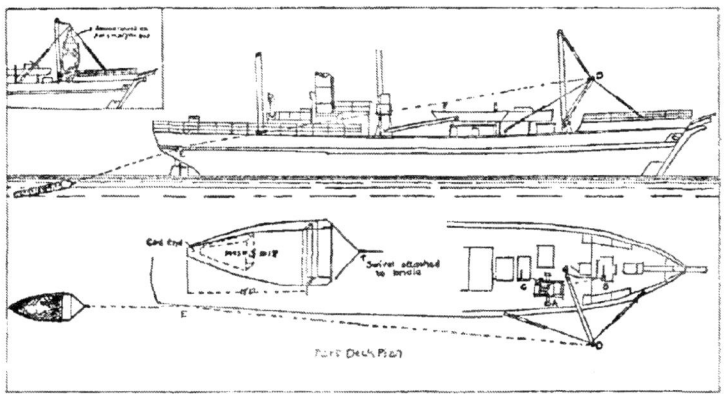

PLAN ILLUSTRATING THE ARRANGEMENTS FOR DEEP-SEA TRAWLING ON BOARD THE AURORA.

A. J. H.

fathoms) was fixed on the port side of the forecastle head. The wire used—.028 of an inch in diameter—was wound in by a belt, worked by a small horizontal engine close by and directly in line with the machine. This engine had been constructed for the "Scotia" (Scottish National Antarctic Expedition, 1902), and was kindly lent to us by Dr. W. S. Bruce. It proved to be just the right thing for winding in the wire. The wire, as it is paid out, passes over a measuring wheel, the revolutions of which record on a dial the number of fathoms out. A spring brake, which is capable of stopping the reel instantly, is kept out of action by the tension of the wire during the run out. When the sinker strikes the bottom, the tension is relaxed and allows the brake to spring back and stop the reel. The depth can then be read off on the dial.

A hollow iron tube called a *driver* is attached to a piece of hemp line spliced into the outer end of the sounding wire. This driver carries one or two weights which become detached when bottom is struck. A sample is recovered in the hollow part of the tube, which is fitted with valves to prevent the water from running through as it is wound up.

The wire when *new* has a tensile strength of 240 lb., and 1,000 fathoms of it weigh 14¼ lb. (in air).

There is invariably a heavy swell in the Southern Ocean, and it will be easily understood that the sudden strain im-

THE MONAGASQUE TRAWL FRAME AND NET.

Photo, Davis.

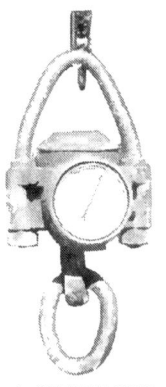

A DYNAMOMETER

A DREDGING BLOCK.

parted to the wire by the rolling of the ship sometimes results in the wire parting as it is being wound in. After some practice, we learned how to handle the vessel and the machine so as to entail the least possible strain on the wire during the operation.

(2) A Kelvin sounding machine for depths up to 300 fathoms was fixed on the poop.

Trawling Equipment.

A short description of our trawling gear may be useful to those engaged in this work on board a vessel not specially designed for it.

The form of trawl used on board the "Aurora" is known as the "Monagasque" trawl. It is of simple construction, and, both sides being similar, it is immaterial which lands on the bottom.

A Manilla swab attached to the side of the trawl is extended to bring up starfish, brittle stars, sea lilies, and other forms of deep sea life.

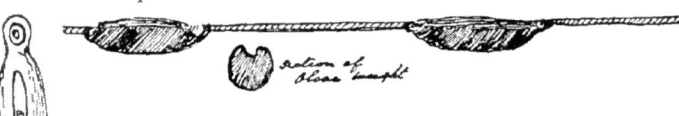

An Olive weight is attached by a rope to the cod-end of the trawl; 28 lb. being generally sufficient. When the net is flowing with its head in good position and streaming neatly aft, an Olive weight of 28 lb. is fixed on the steel trawling wire a few fathoms from the bridle. If trawling in water about 2,000 fathoms deep, a second weight may be put on the wire a little ahead of the first.

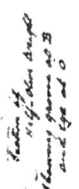

We were provided with 3,000 fathoms of tapered steel wire, varying from $1\frac{3}{4}$ to $1\frac{1}{2}$ inches in circumference, weighing roughly a ton to the thousand fathoms (in air). This was kept on a large reel mounted on standards and controlled by a friction brake. The wire was wound on to this reel by means of a chain drive from the forward cargo winch.

For heaving in, our steam windlass was fitted with a specially constructed drum which absorbed the crushing strain and then allowed the slack wire to be wound on to the reel, which was driven, as nearly as possible, at the same speed. The windlass usually wound in at the rate of 450 fathoms an hour.

A wooden derrick, provided with topping-lift and guys, was mounted on the foremast by means of a band and gooseneck. To the outer end of the derrick, a dynamometer and a 14-inch block were attached. The position of the several parts of the trawling can be seen on the plan, all being designed to stand a strain of 10 tons. In paying out, the wire was led from the derrick-head to a block on the quarter aft constructed to admit of being disengaged from the wire when it was desired to heave in. This block kept the wire clear of the propeller, and allowed the speed of the ship to be regulated while the wire was being paid out.

Trawling in Deep Water.

The vessel is stopped and a sounding obtained; then the derrick is hoisted and the wire rove through the various blocks, the trawl shackled on and the men distributed at their stations. When all is ready, the engines are put at half-speed (3 knots), a course given to the helmsman, and the trawl lowered into the water. When flowing nicely just astern, the order "slack away," is given, the wire being paid out evenly by means of the friction brakes In 1,500 fathoms of water, after the 2,000-fathom mark has passed out, the order is given to "hold on and make fast." Speed is now reduced to $1\frac{1}{2}$ knots and the wire watched until it gives a decided indication of the dragging of the trawl over the bottom. The strain is now taken by the windlass barrel controlled by a screw-brake, backed if necessary by a number of turns round the forward bitts. Dragging slowly over the bottom is generally continued for an hour or longer. The engines are then stopped, and the order "Stand by to heave away," given. This is quickly followed by "Knock out," which means the disengaging of the after-block from the wire; this being done by knocking a pin out of the block, the vessel then swinging round head-on to the wire. "Vast heaving" indicates the appearance of the net at the surface, when the derrick is topped up vertically; the lower end of the net being dragged inboard and the cod-end loosed, allowing the contents to fall on deck.

The biologists usually spent the next twelve hours sorting and bottling the catch.

[Photo, Davis.
DREDGING REEL ON BOARD THE "AURORA."

[Photo, Davis.
THE DREDGING BOOM AND PART OF THE WIRE REEL.

[Photo, Davis.
THE MONAGASQUE TRAWL.

Page 66.

A WINTER CRUISE

The weather during the winter months in the Sub-Antarctic region may be predicted with some certainty. Strong winds, heavy seas, short hours of daylight, and a great deal of fog and gloom! Sounding and trawling under such conditions are difficult operations, and sometimes expensive owing to loss of gear.

We had a fine run through Bass Strait, and were off King Island on May 23rd. The following day the weather was gloomy, and a heavy swell caused a considerable roll. We altered course to the southward. The weather being unfavourable for trawling, we postponed this work and steamed towards the position of Royal Company Islands as given on the French chart.

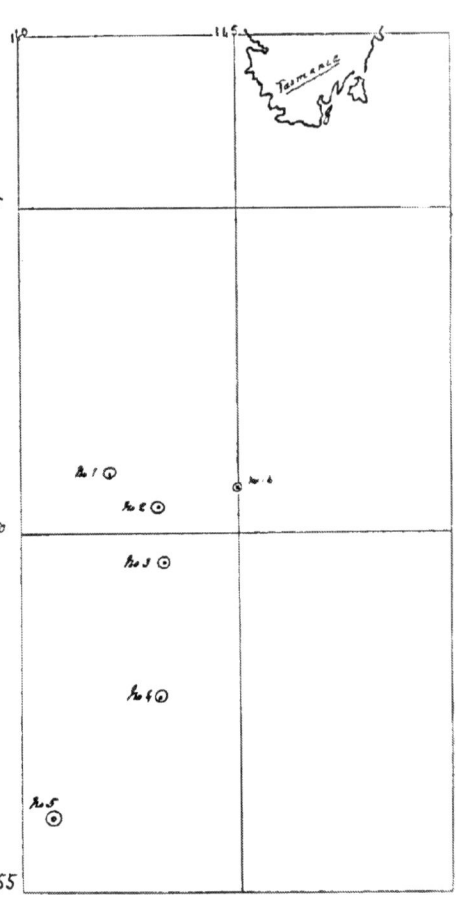

Various charted positions of Royal Company Islands.

The Royal Company Islands.

The Spanish Royal Company (of the Philippine Islands) was incorporated in 1773. About the year 1776, the Spanish ship "Rafaelo" reported the discovery of islands in latitude

49° S., longitude 142° E. These islands appeared on various maps and charts in different latitudes, but on, or very close to, the 143rd Meridian.

Some of the positions are shown on the diagram overleaf, e.g.:—

	Lat S.	Long E.
No 1 Report of Spanish ship	49°	142°
No 2. Arrowsmith's map	49°30'	143°
No 3 Admiralty chart	50°25'	143°
No 4 French chart	52°20'	143°
No 5 Old charts	53°30'	141°
No 6 Old charts	49°10'	145°

Search has been made for these islands by different navigators during the nineteenth century but without success. In May 1909 I sailed over the position given as No. 3 in fine clear weather without seeing any traces of land.*

26.5.12. We took our first sounding to-day in 2,590 fathoms on globigerina ooze. As the moon will be full on May 30th, I am steering direct for the reported position of islands (No. 4). We shall then be able to see what we are about throughout the twenty-four hours, and the search will be more comprehensive.

27.5.12. A cheerless day with fresh, southerly wind and heavy swell. I am afraid our passengers are not enjoying themselves. Mr. Waite is a good sailor and takes matters cheerfully. The others will probably get used to the vagaries of the "Aurora" before we reach Macquarie Island.

28.5.12. The wind has moderated, but there is still a very high sea. We sounded in 2,452 fathoms on globigerina ooze, and a curious accident occurred in connection with this sounding. The two weights (25 lb. each), which we used, remained attached to the driver when we commenced to wind in. We continued heaving up until only 100 fathoms of wire remained out. Then the drum of the Lucas machine burst with a loud report (owing to the extra pressure) We took the last 100 fathoms to the winch and succeeded in recovering weights, driver and sample. Strange to say, the wire did not part, although the ship was rolling heavily. When the drum was examined, a serious flaw was found in the casting.

* See *The Geographical Journal* for December, 1910, page 678

29.5.12. Moderate north-westerly gale. We have managed to fix up our sounding machine by shipping a spare drum.

Good stellar observations this evening enabled us to fix our position (latitude 50° 8¼′ S., longitude 139° 59′ E.).

30.5.12. Although there was a strong north-westerly wind this morning, I decided on sounding at 8 a.m. Bottom was reached at 2,150 fathoms, but as the ship was drifting rapidly to leeward, the wire parted and we lost 2,000 fathoms.

31.5.12. This morning we sailed over the charted position of the Royal Company Islands. There was a heavy sea, but the weather was clear and the moon full. No appearance of *land*!

Observations gave our position at noon as latitude 52°32′ S., longitude 143° 33′ E. We then stood north for four hours, before turning west. The high swell is continuous, and the heavy rolling is a severe strain on the ship.

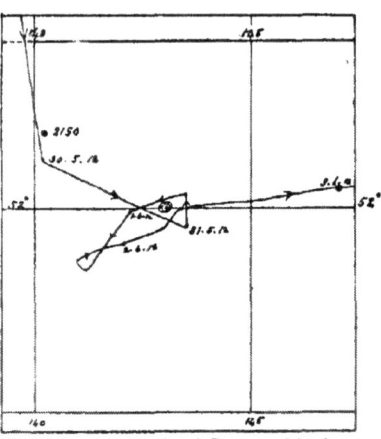

Position assigned to Royal Company Islands on French chart.

The sketch shows the area over which the search was made, but owing to the heavy weather we were unable to take a sounding in the immediate vicinity of the assigned position.

We killed our four sheep to-day, as this boisterous weather has made the animals very unhappy. They could not have survived this exposure for any length of time.

3.6.12. A heavy gale sprang up yesterday, and we are now running before it towards Macquarie Island.

7.6.12. The ship anchored in North-East Bay, in 12 fathoms, about 2.30 p.m. The holding ground was good about a mile from the shore. A moderate gale is blowing from the S.S.W., with occasional showers.

8.6.12. It was still blowing hard at 9 a.m., but we were able to land the mails after a stiff pull ashore. Mr.

Ainsworth's party have built themselves a very comfortable hut, known as "The Shack," and they appear to be enjoying life despite the frequent gales and wintry gloom.

The meteorological station is working well, and the data forwarded daily by wireless to the Federal Weather Bureau at Melbourne and the Dominion Bureau at Wellington, are of more than ordinary value. We climbed the hill to the wireless station where Messrs. Sandell and Sawyer have everything in first-rate order. Two small huts have been built; one for the engine and the other for the transmitting and receiving apparatus. We learned that this remote island has been in communication with the Government Station at Suva (Fiji), and also with Fremantle. Official messages are received from and sent to Sydney. Messages "waved" from the News Agency Station in New South Wales are received in the ordinary course of transmission, and these items are published in the "Macquarie Island Daily News," which is read with the greatest interest by the inhabitants. The aerials have been blown down twice, and the excessive dampness of the climate interferes with the instruments, but I think the greatest credit is due to these two members of the party for the efficient manner in which they have kept the station going under real difficulties during the last six months.

The operators think that messages can be *received* at Commonwealth Bay, although up to the present no message has *come from* that station.

The position of Wild's base has been forwarded to Dr. Mawson from here, in the hope that the message may be picked up by the party in Adelie Land. I was able also to send a message to Professor David announcing our arrival, and informing him that the "Royal Company Islands" were not found in the position indicated on the French chart.

9.6 12. Mr. Blake has practically completed his survey of the northern part of the island. When some soundings have been added, a reliable chart will be available (see Mr. Blake's map, page 120).

Three members of Mr. Ainsworth's party came on board yesterday afternoon. The wind freshened in the evening and has continued too strong to send a boat ashore, so our visitors are in "detention" on board.

G. F. AINSWORTH.
Leader of the Macquarie Island party.

THE MACQUARIE ISLAND PARTY OUTSIDE THE "SHACK."

A low damp mist hanging over the island gives it a gloomy appearance. This bay is a good anchorage in winter, when the wind is not easterly, as, although the wind blows strong, there is very little sea. With two anchors down on good holding ground we are fairly comfortable.

10.6.12. The weather moderated to-day, and we were able to land various stores for the use of Mr. Ainsworth's party. The sealers have a large quantity of oil ready for shipment. Mr. Hatch has just sent a wireless message to say that he is about to send a ship to the island.

Although, during the winter months, gales occur almost daily, the highest velocity of the wind registered ashore since our arrival has been 35 miles.

Mr. Waite and his assistants have been busy collecting specimens for the Canterbury Museum. Several birds have been obtained. We intend moving down the coast to Lusitania Bay, where we can secure some Royal penguins and a Sea elephant.

June 13th–19th We left North-East Bay and followed the coast to Lusitania Bay. There is a curious bed of kelp all down this coast about three cables off-shore We anchored about a quarter of a mile from the land, but as the wind was north-westerly there was not much surf on the beach. Penguins and Sea elephants were visible in large numbers.

Next morning we landed close to one of the old huts. A thick mist crowned the hills, and the low ground was swampy. Mr. Waite arranged to camp in one of the disused huts for a few days.

In case of a south-east wind coming on suddenly, I shifted the anchorage to about a mile from the beach where there was fair holding ground in fifteen fathoms. Next day the weather was so thick that we were unable to see the land, although it was scarcely a mile distant.

Early on the 17th, the wind veered round to north-east, so we put the boat over and brought off Mr. Waite and his party with their specimens. We had a hard pull back to the ship against a strong north-easterly breeze. Very dirty weather for the next twenty-four hours!

We left Lusitania Bay on the 19th, and steered for a position about four miles east of Victoria Point, where we sounded in 900 fathoms, finding *no bottom*—an extraordinary depth so close to the land.

I decided to collect the mail from North-East Bay before putting the trawl over, as the weather was not improving and we might have had to put to sea at very short notice.

20.6.12. We left North-East Bay at 9 a.m. The weather has improved considerably and we hope to have a good day with the trawl before proceeding to the Auckland Islands.

About one-and-a-half miles from the shore we found a depth of 200 fathoms, and at two miles from the shore the depth had increased to 420 fathoms. The engines were then stopped and the net put over, the vessel moving slowly away from the island. After paying out 800 fathoms of wire, the ship drifted away from the trawl

When we commenced to heave in the wire the gear worked fairly well, the only exception being the winding up of the slack wire on the big drum. When the trawl was aboard, the net contained a solitary specimen. The inference was that the net had not been on the bottom. The island slopes so steeply on this side that the ship had drifted into a depth of more than 700 fathoms while the wire was being paid out.

The wind was freshening from the south-west before we had finished, so I returned to North-East Bay to get the gear fixed up again, with sundry small improvements suggested by our experience

The sealers had got the lights out, so we were able to reach an anchorage in North-East Bay at 9 p.m *

21.6.12. We have managed to test the trawling gear, to discover its weak points, and to get these remedied. Our experience yesterday was valuable and cheaply bought, but mid-winter is not the season for trawling in these latitudes

Mr. Waite has secured a good specimen of a Sea elephant for the museum. Four men were required to carry the skin (including the skull and limb bones) from the tussock-grass, across the beach, to the whale-boat. He informed me that the largest animal he had measured was 17 feet long. The sealers stated that the maximum length attained by

---

* See page 117

MR. WAITE OUTSIDE THE "SHACK."

H.M. MAIL ARRIVING AT MACQUARIE ISLAND

[Photo, Masson
THE LANDING AT CAROLINE COVE, MACQUARIE ISLAND.

this animal is 26 feet; this measurement including the hind flippers, which extend beyond the body. The animals are found on the beaches, in the tussocks, or in pools among the rocks; only a few were seen swimming in the sea, though their food is obtained from there. Mr. Hamilton has been investigating the food of seals, and he told Mr. Waite that cuttle-fish enter largely into its composition.

The hind flippers have no forward movement. It is painful to see one of these animals floundering among the rocks in its efforts to reach the sea, where it is perfectly at home.

I may mention that the Sea-elephant is *not* " on the verge of extinction." On the contrary, the sealers who gain their living at Macquarie Island by rendering down sea-elephants for their oil rely entirely for their supply on those animals which come ashore in the vicinity of the digesters.

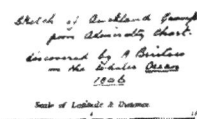

The Auckland Group.

CHAPTER IX

A WINTER CRUISE (*Continued*)

THE AUCKLAND ISLANDS

22.6.12. We sailed from North-East Bay at 8 a m and obtained a good line of soundings to the north-east until the Judge and Clerk were abeam and about a quarter of a mile distant. These rocks were about 8 miles from the wireless station. A sounding gave at this point $\frac{}{100}$ fathoms.* A course was then set (N. 46° E. true) for the Auckland Islands.

To-day being mid-winter day will be celebrated at the Bases in Antarctica with all due ceremony.

The Auckland Group consists of one main island and several smaller ones, separated by narrow channels. They lie in the track of homeward bound vessels from Australia via Cape Horn. The group was discovered in 1806 by Captain Bristow in the "Ocean," owned by Samuel Enderby. There are two spacious harbours; a northern, now called Port Ross, and a southern, Carnley Harbour The position is about 340 miles north-east from Macquarie Island and 180 miles south of Stewart Island, New Zealand. The total area of the group is about 200 square miles, and the formation is volcanic; chiefly basalt and greenstone. Rocky cliffs from 100 to 750 feet in height extend along the western coast with deep water right up to the shore. The eastern coast is much indented, and the inlets are well sheltered from the northwesterly winds.

Enderby Island forms the northern boundary and South Island the southern. From the shores of the harbours, low forest covers rising ground followed by a belt of underwood, above which grassy slopes rise to the summits of hills The trees are stunted owing to the violent westerly gales, and their bent growth shelters a luxuriant crop of ferns and flowering

* NOTE—$\frac{}{100}$ indicates bottom not reached at 100 fathoms.

plants. Much of the surface is covered with peat, which in some places is several feet in depth.

Sea-lions are numerous at certain seasons of the year. A study—from personal observations—of the habits of this seal is given in the appendix to *Musgrave's Journal*.* Captain Musgrave, with his mate and three men, spent eighteen months on the main island after the wreck of the schooner " Grafton " in Carnley Harbour in January, 1864.

Heavy weather prevailed during our passage to the Auckland Islands. Several small but interesting fishes were washed up on deck during the voyage.

24.6.12. Land was sighted at 9.30 a.m., and we entered Carnley Harbour at 3 p.m. We anchored off Flagstaff Point in twenty fathoms, as darkness was coming on and our chart was unreliable. There was a strong breeze from W.N.W., but we lay comfortably until the following morning.

The eastern entrance to Carnley Harbour is formed by two bluff points about two miles apart. The distance to the head of the harbour is about fifteen miles.

25.6.12 I decided to look for an anchorage with less water, so we stood up to Figure of Eight Island and found good holding ground (in 9½ fathoms) in a land-locked bay to the north of this island. We all felt glad of the rest afforded by this position. In the afternoon we got the launch over and took some soundings. The mud extends well out from the head of bay with only two fathoms of water. The small dredge was tried and a few specimens were obtained. To-morrow we intend to take the launch as far as Camp Cove to inspect the provision depôt, maintained there by the New Zealand Government for the use of unfortunate castaways.

26.6.12. The wind had moderated during the night, so the launch soon covered the distance of five miles to Camp Cove. We landed near the two white-painted sheds which form the depôt. The larger shed contained stores of biscuits, tinned food, matches and other necessaries; in fact, everything required to support a large party. The smaller hut was empty, but on the outside were carved many names of mariners from the crews of shipwrecked vessels. The name of a French ship " Anjou " was noticed as being one

* " Cast away on the Auckland Islands," T. Musgrave Edited by J. J. Shillinglaw Lockwood & Co , London, 1866

THE "AURORA" AT ANCHOR OFF EREBUS AND TERROR COVES, AUCKLAND ISLANDS.

DEPÔT FOR SHIPWRECKED MARINERS AT CAMP COVE, AUCKLAND ISLANDS.

of the tragic wrecks on this primitive record. A good boat was housed at a short distance from the sheds.

The New Zealand Government's steamer "Amukura" had visited the depôt in November, 1911, and a notice of this visit was posted up in a conspicuous position. These visits are paid usually about twice in the year. A note recording our visit was left.

We next visited the camp kitchen left by the Sub-Antarctic expedition of 1907. We soon had a fire going and enjoyed a good lunch under the shelter of the old kitchen. The weather was very chilly; a thick mist rolling down from the surrounding hills. Before returning to the ship the light improved, and we secured some photographs.

28.6.12 Raining all day yesterday.

This morning we made a trip in the launch to various points round the bay with a view to making a rough plan of the anchorage; which I consider an excellent one, being easy of access, having good holding ground, and being completely land-locked. This afternoon we got the launch on board as we leave for Port Ross to-morrow.

After clearing the entrance to Carnley Harbour we found a heavy sea outside. There was a fresh southerly gale with frequent snow-squalls. The ship proceeded up the eastern coast at a good rate, keeping well out. We passed Blanche Rock, which appeared to be an extensive patch, as the sea was breaking heavily all round it.

29.6.12. After clearing the low reef which extends for some distance beyond Green Island, we stood in for Port Ross, which is a good harbour, but not to be compared with our recent anchorage. We brought up in Erebus Cove (in $12\frac{1}{2}$ fathoms) just west of Shoe Island. There was no sign of any other vessel being in the port unless one should be at the head of Laurie Cove. We can look in there to-morrow.

30.6.12 A fine day with a light south-westerly breeze.

We visited the depôt at Erebus Cove. It consists of three large sheds, the largest being fitted with bunks, and a fireplace. Another is well stocked with provisions, and the third contains a good boat, everything was in good condition. We spent some time examining the various inscriptions carved by castaways on the walls of the huts. A full list of the crew of the "Dundonald" (2,205 tons), which had been wrecked on Disappointment Island, had been worked with a nail on a piece of tin, by one of the sixteen survivors. The story of the loss of this vessel, the hardships endured by the survivors, and their relief by the New Zealand steamer "Hinemoa," in 1907, is one of thrilling interest.

We next visited the cemetery, where four crosses mark the respective resting places of four persons, one being that of J. Peters, mate of the "Dundonald," who died from exposure after the wreck. These white crosses, surrounded by ferns, served to remind us of the sufferings endured by those who have been "cast away" on such remote islands.

We looked into Laurie Harbour, but there was no vessel there. It is better sheltered from the sea than Erebus Cove, being nearly land-locked. I searched for any traces of the Enderby settlement which had been established on the southern side of Erebus Cove in 1851, but could find none. The site of the buildings is now overgrown with rank vegetation.*

We shifted our anchorage to Terror Cove; a pretty little bay with low hills covered with bush on three sides. A stream of good water runs into the cove.

2.7.12. Fine clear weather and bright sunshine combined to give us the finest day we have had since leaving Bass Strait. We visited Observation Point, marked by a cairn, where Captain J. Ross set up his observatory in 1840. Alongside, there is a flat stone marked "German Expedition, 1874."

I noticed a shoal patch, carrying two fathoms, off Observation Point. This was not marked on the chart. In the afternoon we landed on Rose Island, close to the place where a boat is harboured. I walked across the island through the bush, and got a good view of the northern coast. There are some fine basaltic columns, eighty feet high; the bases being often weathered out into deep caves.

A variety of sea birds frequents the Auckland group. Some are similar to those found on Macquarie Island, but the flightless duck,† of which Mr. Waite obtained six specimens to-day, is confined to certain islands of the Auckland

---

* The settlement was abandoned in 1852, all the plant and material being removed to New Zealand.

† Nesonetta Aucklandica

RATA TREES AT EREBUS COVE.

SHIPWRECKED MARINERS' DEPÔT, CAMP COVE.

[Photo, Gillies.
THE "AURORA" AT ANCHOR, CARNLEY HARBOUR.

[Photo, Gillies.
THE "AURORA" AT ANCHOR OFF SHOE ISLAND.

[Photo, Gillies.
THE LAUNCH AT CAMP COVE.

THE AUCKLAND ISLANDS.

A WINTER CRUISE

group. These birds live among the kelp and are not easily found, except when seen sitting in pairs on the rocks. They can fly a few yards, but appear to use their wings during the breeding season only, their nests being constructed about twelve feet above the level of the water.

Various kinds of petrels are found on Shoe Island, where the turf was riddled in all directions by their burrows.

5.7.12. We anchored in Sandy Bay, Enderby Island. There is a good landing place close to the depôt, where stores and a boat are kept. Among the stores I noticed a venesta case marked " S. Y. Nimrod," which contained dried vegetables and was evidently one of the cases which had been sold on the return of the Antarctic Expedition (1907—09), to the New Zealand Government

We crossed the island to the northern coast. After passing through some bush we came to the level top covered with short grass in which were a few swampy patches. The seaward face of the northern coast consists of basaltic cliffs from eighty to one hundred feet above the sea, which, during heavy gales, must reach the top of the cliffs, as a large piece of fresh seaweed was found in one place close to the edge.

We could see Bristow's Rock, which "breaks" about once in ten minutes, lying some three miles off in a northerly direction. In thick weather a ship would probably strike before seeing it. Mr. Waite bagged twenty-five silver-grey rabbits, which abound on the island, and a new variety of penguin was secured on our way back to the beach. Altogether we spent an interesting day

On July 6th we left for New Zealand and had moderate weather during the run to Port Lyttelton Five soundings were taken. The trawl was put over in 345 fathoms on July 9th, but the net fouled on a rocky bottom, so we gained nothing but experience by this operation.

Although the winter cruise of the "Aurora" was not fruitful in deep-sea work—mainly owing to the very stormy weather—it afforded some necessary training for officers and men in the handling of deep-sea gear. At a later date we were able to realize how much we had learned during our first cruise in the Sub-Antarctic region.

Port Lyttelton was reached on July 11th, and we received a hearty welcome from the people of Christchurch. Mr.

Kinsey gave us the most valuable assistance during our stay. We proceeded to Melbourne, and, after a good passage, anchored off Williamstown on August 17th.

While the " Aurora " was refitting in the dockyard at Williamstown, I had the privilege of enjoying a fortnight's cruise on board the Commonwealth Fisheries Investigation vessel " Endeavour,"* and I shall always feel grateful for this exceptional opportunity of seeing the working of deep-sea appliances, under the direction of Mr. Dannevig, on board that vessel. Different patterns of nets and a variety of deep-sea apparatus were used during the trip. The voyage was not only very instructive, but also most enjoyable.

* *The loss of the s.s. " Endeavour."*—The vessel left Macquarie Island on December 3rd, 1914, for Hobart. She carried a crew of 19, all told, with Mr. H. C. Dannevig, the Director of Commonwealth Fisheries, and his staff. Mr. C. T. Harrison had filled the post of biologist on board since his return from Antarctica, where his excellent work won special mention from the leader of the Australasian Antarctic Expedition. The vessel has not been heard of since her departure from Macquarie Island. Three large steamers spent several weeks in searching for any trace of the missing vessel, but without success. The only conclusion is that she must have foundered on her passage to Hobart.

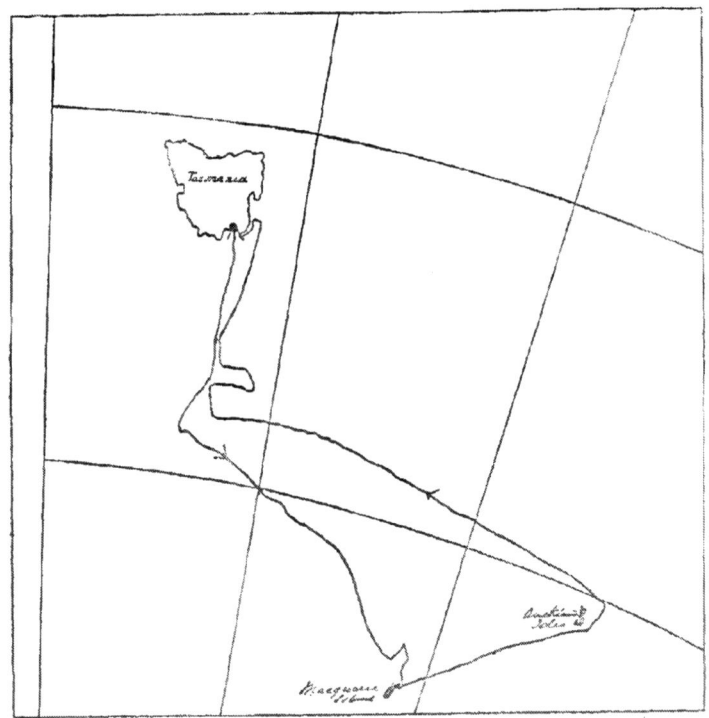

Track of the "Aurora." Spring Cruise, 1912.

CHAPTER X

THE SPRING CRUISE IN THE SUB-ANTARCTIC

NOVEMBER 12TH TO DECEMBER 14TH, 1912

> It ain't the individual, nor the Army as a whole
> But the everlasting team work of every blooming soul.
> *Kipling,* "Teamwork."

THE " Aurora " left Hobart on November 12th for a month's cruise in the deeper waters between Tasmania and the Auckland Isles. Sounding and trawling were undertaken as opportunity offered, and weather permitted. Professor Flynn, of Hobart University, accompanied us as biologist.

The longer hours of daylight at this season, and sundry improvements in our deep-sea apparatus, made me hopeful of fair results during the limited time at our disposal, as we had to be back at Hobart not later than December 15th to prepare for a voyage to Commonwealth Bay. Our programme included a call at Macquarie Island to land stores and mails. So we steered southward from Hobart across a tract of ocean over which no soundings were shown on the chart, expecting to find an average depth of about 2,000 fathoms when clear of the continental shelf.

Our first sounding was taken at 8 a.m. on November 13th about 80 miles S.S.E. from Hobart, giving a depth of 658 fathoms on a hard bottom with some shells. At noon the depth was 1,475 fathoms on mud.

On the 14th, at 8.45 a.m., the depth was 2,083 fathoms on mud. At 6 p.m. the depth recorded was 1,940 fathoms on mud.

On the 15th, at daylight, the water had shoaled from 1,940 fathoms to 792 fathoms (on hard bottom). The sounding was repeated and a depth of 794 fathoms obtained, still on hard bottom. This was a very rapid rise in a distance of about 50 miles. As the depths are marked on the track of the "Aurora" from Hobart to Macquarie, shown on the opposite page, repetition is needless.

From the afternoon of the 15th, a fresh westerly gale with high seas caused the vessel to labour heavily and stopped sounding operations for forty-eight hours

At 5 p m. on the 17th, we paid out 1,570 fathoms of sounding wire without finding bottom. This showed we were getting into deeper water again, so the course was altered to south-east.

Proceeding towards Macquarie Island, soundings were taken when the weather permitted, all being over 2,000 fathoms, until November 21st. At 7 p.m. on that date, a depth of 1,405 fathoms was found in latitude 53° 46′ S , longitude 158° 46′ E. As the weather was unusually fine for this locality, the trawl was put over and towed slowly in a north-easterly direction; 1,900 fathoms of wire being paid out.

The trawl apparently fouled the bottom, as the strain on the dynamometer suddenly increased to 7 tons. After some time we got the trawl clear of the bottom, and the strain was reduced to $2\frac{1}{2}$ tons. When we got the trawl aboard there was ONE fish in the net, which was torn, the frames being badly bent. We then sounded in 636 fathoms on hard bottom. It appeared that the trawl had

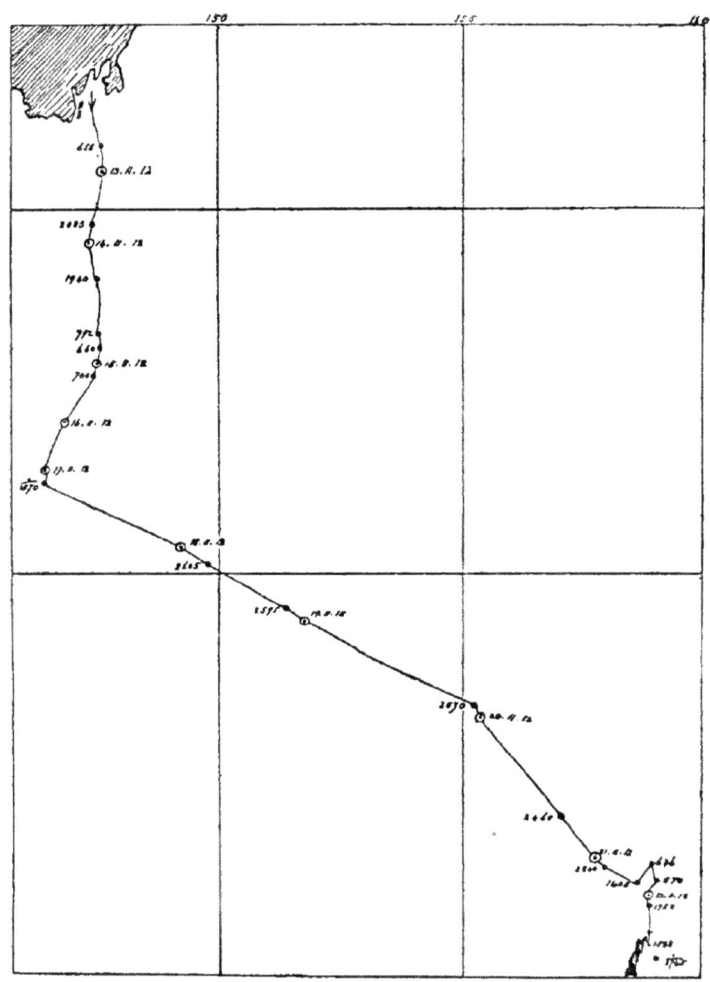

Track of the "Aurora" from Hobart to Macquarie Island.

been put over near the foot of a steep rise, and then had been towed UP into comparatively shallow water.

At this point we steered towards Macquarie Island. About thirty miles north of the island the water deepened to 1,750 fathoms. At 6.15 p.m. the "Judge and Clerk" rocks were abeam. Although these rocks are not more than eight miles from North-East Bay, the outline of the island could not be seen; a dense mist covering the land rendered it invisible from seaward.

At 7.30 p.m. we anchored in North-East Bay. The Island Party came off in their boat, bringing us items of news from

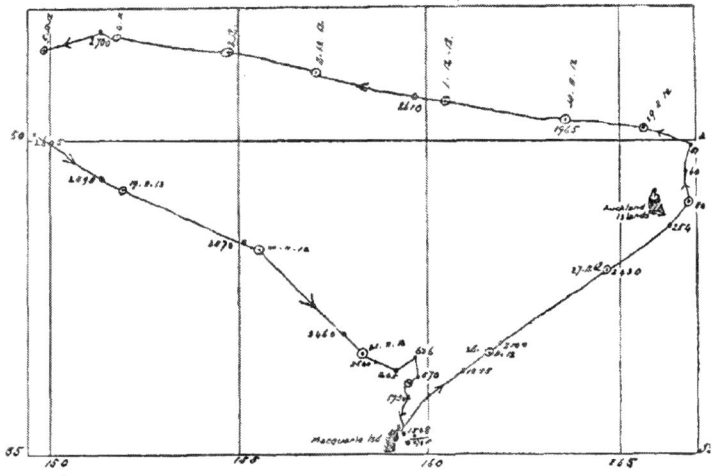

Deep sea soundings.

various parts of the globe. The arrival of the "Aurora" was "waved" to Australia and to New Zealand before midnight.

23.11.12. The weather being favourable, a number of soundings were taken round the northern part of the island; the depth varying from 11 to 398 fathoms—the latter depth was found about three-quarters of a mile east of the Judge and Clerk.

24.11.12. About three miles east of North-East Bay we found a depth of 1,548 fathoms. At ten and a half miles east of the middle of the island 2,745 fathoms were paid out without finding bottom. This was the greatest depth found

SOUNDING OFF MACQUARIE ISLAND WITH KELVIN MACHINE.

THE SPRING CRUISE IN THE SUB-ANTARCTIC 79

during the cruise. Unfortunately, this was all the wire on the machine at the time.

There are no soundings on the chart between Macquarie Island and the Auckland group. Many people have expressed the opinion that shallow water would be found in this direction.

The sketch shows the nature of the sea-floor sailed over after leaving Macquarie Island, until December 5th, when we were approaching the bank met with about 200 miles south of Tasmania on the outward voyage.

5.12.12. At 6 p.m. we sounded in 1,076 fathoms. This showed we were approaching the Rise, so a westerly course was followed in the hope of outlining this feature in that direction.

6.12.12. To-day we found 1,300 fathoms at a point which appeared to be near the southern end of the bank Then a westerly gale obliged us to keep away to avoid damage to the vessel. Three soundings were taken on hard bottom.

7.12.12. At 6 a.m. the "Aurora" was on an easterly course ; we were fortunate in securing three soundings.

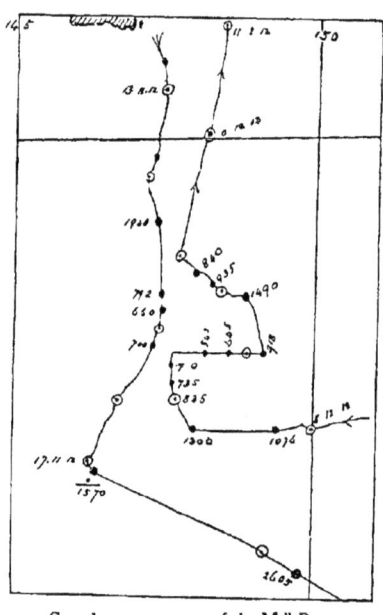

Soundings in vicinity of the Mill Rise

8.12.12. We were again fortunate in obtaining three soundings, under adverse conditions, to the west and northwest.

9.12.12. The weather has been so bad and the sea so heavy that we have been unable to obtain soundings. I have decided to steer for the east coast of Tasmania near Maria Island, where we may be able to do some trawling in the smoother waters under the lee of the land This last gale has been blowing for four days, with only very short intervals of moderate weather.

12.12.12 We are all pleased to get under the shelter of the land. The trawl was put over in 1,300 fathoms about 25 miles from the land; the result was unsatisfactory. We then stood in-shore and shot the trawl in 75 fathoms. After trawling for about an hour, the net came up FULL, which compensated us for the disappointment in the earlier part of the day.

13.12.12. We shot the trawl in 1,300 fathoms, and, after towing it for two hours, we recovered the net containing a large octopus and several specimens of marine life. The bottom was good, as the net was not chafed at all.

14.12.12. We reached Hobart just before noon, and commenced preparations for a voyage to the Antarctic.

Considering the limited time at our disposal, which was further curtailed by the occurrence of severe gales, the results of this cruise were satisfactory.

Learning to handle deep-sea apparatus, especially trawling gear, is rather disheartening work, but earnest endeavour will overcome many obstacles.

The determined efforts of officers and men " to do their best " in carrying out this portion of the work of the Expedition assigned to the "Aurora" was rewarded by the proficiency attained at a later date, notably in 1914. During this recent cruise the trawl had been put over on five occasions, but on only two of these were the results successful.

The delineation of the sea-floor, within the area shown on the sketch map, gave some results of more than usual interest :

(1) A submerged bank was discovered about 200 miles to the southward of Tasmania, covering an area estimated at 15,000 square miles. The least depth of water so far ascertained over this rocky plateau is 543 fathoms. This may be a remnant of the old land-bridge which is supposed to have connected Tasmania with Antarctica. The discovery of this bank will doubtless form a subject for discussion by scientists. It has been named Mill Rise.*

(2) The contour of the ocean floor has been outlined

* So named by Dr. Mawson after Dr H. R. Mill, whose interest in Antarctic Exploration is so widely known.

over an area where depths had not been shown on the latest Admiralty chart (1911).

(3) A rocky bank was discovered about 60 miles north of Macquarie Island rising to within 570 fathoms of the surface. The surrounding depths are shown.

(4) A trough depression was proved to exist between Macquarie Island and the Auckland group.

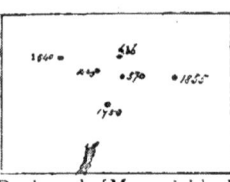
Depths north of Macquarie Island.

The number of soundings taken in the vicinity of Mill Rise was too small to give more than a rough idea of the submarine surface. We could only indicate the direction in which the work might be extended by a powerful vessel specially equipped for such investigations, and with time at her disposal.

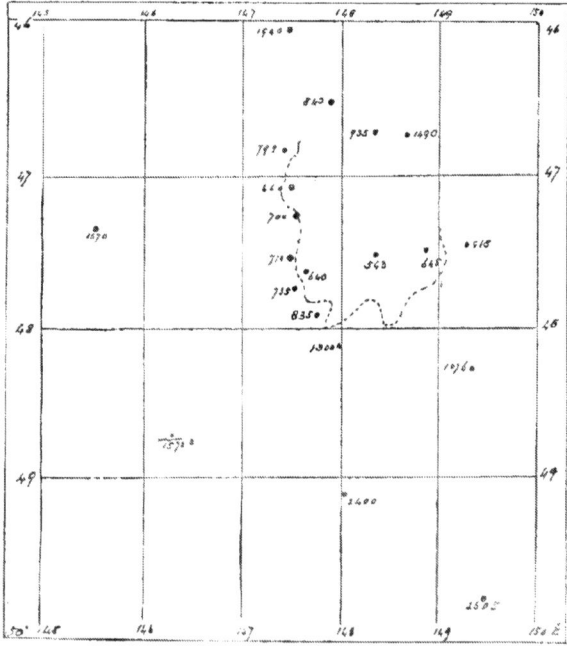
Approximate sketch of the soundings taken on the submarine bank called Mill Rise, and the vicinity, by the "Aurora" in 1912.

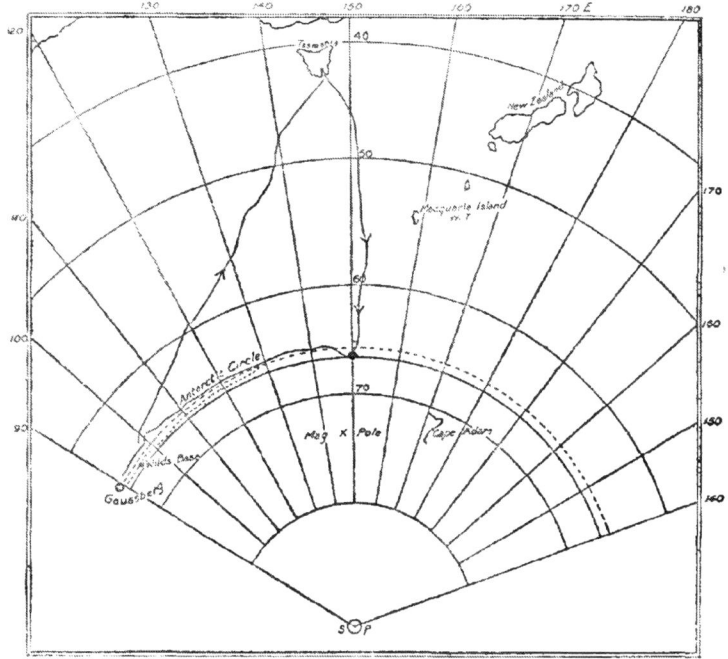

Second Antarctic Voyage. Track outward and homeward are shown approximately.

CHAPTER XI

THE SECOND ANTARCTIC VOYAGE

December, 1912 to March, 1913

Hobart, December 24th, 1912. The preparations for our second voyage to the Antarctic bases were completed to-day. We leave here for Commonwealth Bay on the 26th inst.

Captain J. Davis, of Hobart, will accompany us as whaling expert, Mr. Jeffryes as wireless operator, Mr. Van der Gracht as marine artist. With Mr. Eitel, secretary of the Expedition, the number of our party on board will be twenty-eight in all.

We have a heavy mail of postal packages, twenty-one dogs, presented to the Expedition by Captain Amundsen, thirty-five sheep, and 521 tons of coal.

We have been working at high pressure for the last ten days, so Christmas Day ashore will be a well-earned holiday for the ship's company. I have arranged that we shall dine together at the Orient Hotel and toast our comrades in the far-off regions of the new " South Australia."

26.12.12. Our departure had been fixed for 10 a.m. sharp. I wish to place on record that, notwithstanding the festive season and the proverbial hospitality of the citizens of Hobart, every man was aboard and ready for duty before the appointed hour. As we drew away from the pier, amid the cheers of those who had come to wish us Godspeed, the weather was simply perfect, and the scene on the Derwent bright and sunny. Captain J. Davis acted as pilot. We stopped at the Quarantine Station to embark the dogs and, by noon, we were steaming down Storm Bay. . . . About 11 p m. Pedro Blanca Rocks were abeam, while a fresh westerly wind was blowing as we set a course to the south-west. On arriving outside the Bay, the ship rose to a heavy swell.

For some days after leaving Storm Bay, heavy weather prevailed. On the 29th a good sounding was obtained in spite of adverse conditions.

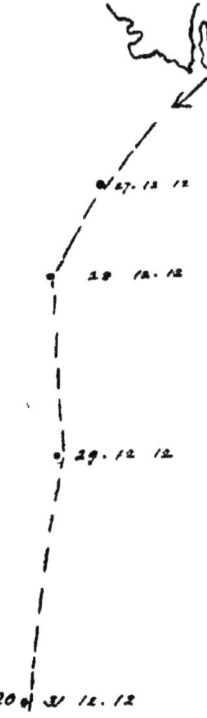

Track southward

On the 31st, part of the rail on the forecastle head was carried away by a heavy sea. Being in the vicinity of the Royal Company Islands, we managed to take a sounding with great difficulty, and obtained 2,020 fathoms on ooze.

31.12.12. The wireless gear was put up to-day, the aerial wires being stretched between the fore and mizen masts.

January 1st, 1913 The wind is moderating, although

a high sea is still running. We sounded in 2,170 fathoms, but the wire parted when heaving in.

3.1.13. At noon we were about 470 miles west of Macquarie Island. Sunset was visible to-night for the first time since leaving Tasmania.

4.1.13. Fresh westerly breeze with occasional snow-squalls. The wireless operator was able to hear H.M.S. "Drake" at Hobart and also the station at Macquarie Island.

5.1.13. Clear sky and bright sun with only a moderate swell; the first really fine day since December 26th. Numerous "Blue Billies" or prions flew about the ship all day. Two good soundings were obtained—both in 1,900 fathoms.

6.1.13. Hard gale with heavy snow-squalls all day.

7.1.13. Fresh N.N.E. breeze, with some rain. We stopped for six hours, to effect necessary repairs in the engine-room.

8.1.13. Sounded in 2,250 fathoms; the sample of mud contained a small black nodule.

We sighted a floating cask and got it aboard. It was a ship's oil cask, empty, and gave no indication whence it had come to be picked up in latitude 62½° South.

9.1.13. Moderate wind, but confused sea. The weather is unsettled; clear intervals alternating with thick fog.

Track southward.

10.1.13. A moderate south-east gale is raging, with a good deal of snow. We sighted the first ice about 6 p.m., the depth of the water still exceeds 2,000 fathoms.

11.1.13. A very fine, calm day. The welcome change

was fully appreciated. At noon the ship met loose pack, with a strong blink of heavier pack to the south. We avoided entering the ice by steering to the south-west.

At 7 p.m. we were able to steer west in clear water, with a few bergs in sight and a strong blink to southward.

Several whales have been round the ship to-day. Captain James Davis tells me these are blue whales, known to whalers as "sulphur bottoms."

About midnight, the course was altered to the south-west, with the intention of coming up with the barrier met with in 1912 well to the eastward.

12.1.13. At 4 a.m. we sounded in 350 fathoms on small stones. At 6 a.m. a sounding gave 184 fathoms on mud and small stones.

At 10.30 a.m. the depth was 230 fathoms on hard bottom. We have sailed right over the position of the barrier, met with in 1912, and seen no trace of it. Apparently Côte Clarie, described by d'Urville in 1840, and discovered by us to have disappeared in 1912, was a similar ice-formation.

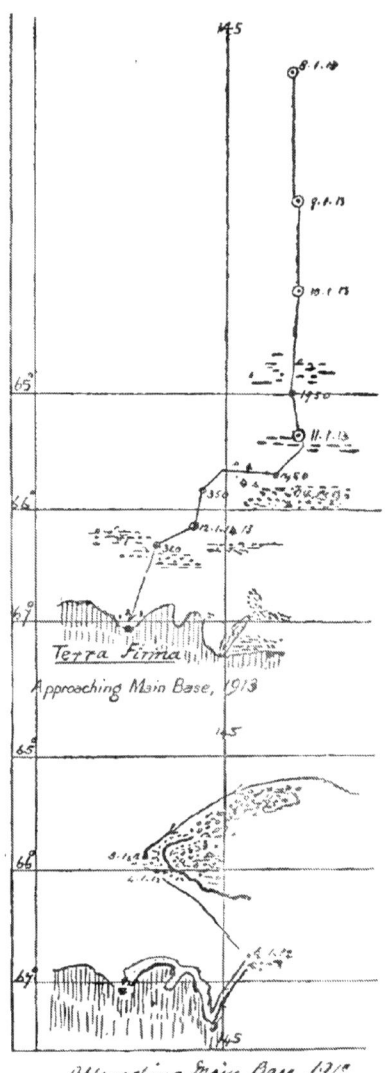

Approaching Main Base, 1913

Approaching Main Base, 1913

Some heavy bay-ice was met during the forenoon, and we steered west hoping to clear it. At noon our position was in latitude 66° 14′ S., longitude 143° 37′ E. An hour later we stopped for sounding and found a depth of 320 fathoms on ooze. As the ice appeared to be extending for some distance to the westward, the ship altered course to the southward and worked through floes and bits of bergs until 5 p.m. when open water was reached and the blink of the land was visible. The wind was freshening from the S.S.E. as we approached the land, which was plainly visible at 9 p.m. A course was set for the Main Base and we were off the Mackellar* Islands at midnight. The weather was squally.

13.1.13. We passed between the Islands and a reef to the west, through a passage carrying from fifteen to twenty fathoms of water. At 2 a.m. we dropped the starboard anchor in twenty-five fathoms, with ninety fathoms of chain.

At 6 a.m. the ship was struck by a fierce squall; the Mate

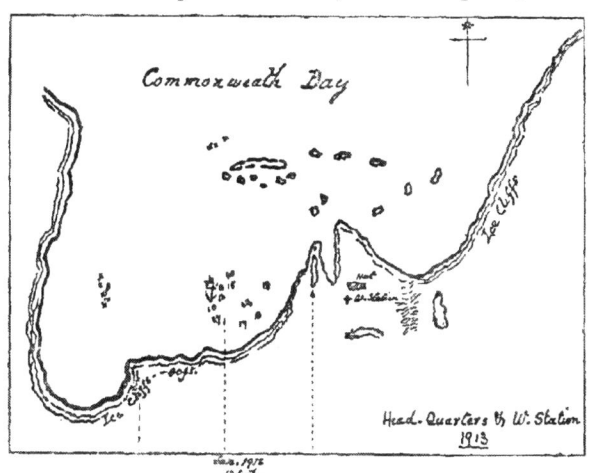

The Anchorage, Commonwealth Bay.

reporting immediately after that the heavy relieving-tackle on the cable to which the ship was riding had carried away. This caused the chain to run out, and, before anything could be done, our best ground tackle had gone overboard.

* Named by Dr. Mawson after Mr. C. D. Mackellar, a staunch supporter of the Expedition from its inception.

MAIN BASE.
The Boat Harbour and Landing Place.
Photo, Hurley.

MAIN BASE, 1913.
Photo, Correll.

At 7 a.m. the port anchor was dropped in ten fathoms about eight hundred yards west of the first anchorage, with sixty fathoms of chain.

At 9 a.m. the wind shifted suddenly to the north and the "Aurora" swung in-shore, but the depth under her stern proved to be over seventeen fathoms. After a few puffs from the north the wind shifted back to south-east, and by noon had moderated sufficiently to allow of preparations being made for going ashore

We got the launch over and took with us the mail and sundry packages of refreshing fruit from the Home Land. The boat-harbour was reached before any one of the shore party had seen the "Aurora."

We received a hearty greeting from nine * wild-looking men, some with beards, bleached by the weather, but all in first-rate condition after the severe winter and the later sledging journeys. The excitement caused by the sight of the mail-packages can be better imagined than described. The first news from home, after twelve months interval !

We learned that five sledging parties had left the Main Base in November, 1912. Of these, *two* had returned all well, the return of the other three parties was expected about January 15th

Commonwealth Bay had proved to be a terribly windy station. The average hourly velocity had been far beyond anything previously recorded. The charts of the self-recording instruments had shown the average for 1912 to have been fifty miles per hour. Hourly velocities of over one hundred miles had been recorded, while twenty-four hourly averages of over ninety miles had been registered. One of the wireless masts had been blown down during a hurricane gust on October 13th.

During the winter months, the drifting snow had never ceased, and, even in the height of summer, blizzard followed blizzard in rapid succession. The nature of the prevalent weather at Winter Quarters, from March to October, 1912, may be gathered from the following extracts, contrasting a

---

* The nine members were: Murphy, Bage, Close, Hannam, Stillwell, Laseron, Hurley, Webb, and Hunter.

hurricane in the summer and winter; the air-temperature registering as low as −28° F. (in winter).

*" Picture drifting snow so dense that daylight comes through dully, though maybe the sun shines in a cloudless sky; the drift is hurled screaming through space at 100 miles an hour, and the temperature is below zero, Fahrenheit."

" Shroud the infuriated elements in the darkness of a Polar night and the blizzard is represented in a severer aspect. A plunge into the writhing storm-whirl stamps upon the senses an indelible and awful impression, seldom equalled in the whole gamut of natural experience. The world a void, grisly, fierce, and appalling. We stumble and struggle through the Stygian gloom; the merciless blast stabs, buffets, and freezes; the stinging drift blinds and chokes. In the ruthless grip of the blizzard we realize that we are—

" poor windlestraw,
On the great sullen, roaring pool of Time "

Tents, similar to those used successfully in the Ross Sea area were soon reduced to shreds Three parties, provided with such tents, had left the Hut in September to reconnoitre the immediate vicinity. The tents, which had been strengthened, were soon torn to ribbons, and the men were fortunate in being able to regain the Hut without serious accident, although Madigan, Whetter and Close were badly frostbitten. Two underground rooms were excavated in the glacier-ice on the rising plateau at distances of 5½ and 11¾ miles respectively from the Hut. In these ice-caves, stores were accumulated during the spring, in readiness for the summer sledging.

The 5½-mile depôt was known as " Aladdin's Cave."

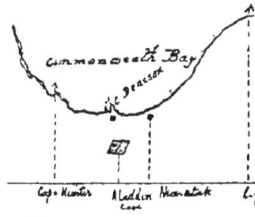

The launch made some trips to the ship during the afternoon. I returned on board and had a look at the cable. Although the weather was fine, changes were apt to occur with short warning At midnight it was blowing hard from the south-east and the launch was hauled up in

* *The Home of the Blizzard*, Sir Douglas Mawson W Heinemann 1915.

the davits. The chain continued to hold well, although there were violent squalls at intervals.

14.1.13. We had a good specimen of "summer weather" to-day—clear sky and bright sun, with a strong gale from the S.S.E. Communication with shore was necessarily suspended.

15.1.13. Mr. Murphy came off this morning in the launch and furnished me with some particulars about the sledging parties still out. Dr. Mawson with Mertz and Ninnis had gone to the south-east. The party was well equipped and provisioned, and they had taken eighteen dogs for haulage. They had been away for sixty-six days; but Murphy did not consider that there was any cause for anxiety, as he believed that all three parties would reach the Hut within forty-eight hours, and that, in all probability, bad weather was the cause of delay.

Bickerton, Hodgeman and Whetter had been out for forty-three days, in a westerly direction, while Madigan, Mclean and Correll had been away for seventy days in an easterly direction.

16.1.13 The Eastern Party (under Madigan) returned to-day. We were glad to greet our comrades, who were all well, on their return from a successful journey over the sea-ice as far as longitude 150° 21′ E

Dr. Mawson had left a letter for me, with instructions to take charge, in case he had not returned by January 15th. After a consultation with Madigan and Bage as to the position of affairs, owing to the non-return of two parties, we decided to have the wireless gear put into working order IMMEDIATELY, as one of the masts had been blown down during a gale, and the re-erection of this mast would take several days to complete.

18.1.13. The Western Party returned to-day The air-tractor sledge had been told off to assist them, but it broke down about ten miles south of the Hut. The party had travelled over the highlands of that part of the coast seen by d'Urville, coming close to the cliffs in longitude 138° E, whence frozen sea was seen to the west.

The erection of the wireless mast has been commenced. Captain James Davis and the Chief Officer, Fletcher, have been away in the launch dragging for the cable lost on the morning we arrived; the grapnel has been buoyed until operations can be resumed to-morrow.

19.1.13. The weather was calm this morning, and the day has been spent in trying to get hold of the chain, with the valuable assistance of Captain James Davis, who worked as hard as any of the salvage crew. All efforts to get hold of the chain were unsuccessful.

On January 20th all the members of the shore-party were assembled at Winter Quarters with the exception of the Leader and his two comrades. The general opinion was that the party had been delayed by bad weather or by one of the minor difficulties inseparable from a long journey over the plateau. But every member had implicit confidence in the ability of the Leader to surmount obstacles.

On thinking over the position, I came to the conclusion that Dr. Mawson, having laid his plans and fixed on January 15th as the latest date of his return to Winter Quarters, would have carried out his expressed intention, unless something had occurred of a serious nature. The party was now FIVE DAYS overdue. The "Aurora" would have to sail westward, to relieve Mr. Wild's party, before the season was too far advanced, or, not later than January 30th.

I decided to have a provisional notice posted up in the Hut, stating that the non-arrival of Dr. Mawson's party rendered it necessary to prepare for the establishment of a Relief Expedition to remain in Antarctica for another year, if necessary; and, appointing Messrs Bage, Bickerton, Hodgeman, Jeffryes, and Dr. Mclean members of such a party, under the command of Mr. C. Madigan.

The remaining members of the Expedition, who would be leaving by the "Aurora," were instructed to remain on shore until the vessel was ready to sail, and to render every assistance in the landing of the stores, for the use of the Relief Expedition, during the brief intervals of fair weather when communication between ship and shore was possible.

21.1.13. The fine weather of yesterday and to-day has greatly facilitated the erection of the mast; good progress is being made. I have arranged with Mr. Madigan to have the provisional notice posted up in the Hut to-morrow. He will send a search party to examine the vicinity of Aladdin's Cave for any signs of the missing members.

A LARGE BERG. *Photo, Gillies.*

LOOKING NORTH IN THE VICINITY OF MAIN BASE. *Photo, Correll.*

22 1.13. This is the seventh day of anxious waiting and hoping to welcome the absentees!

I went ashore this evening and inspected the work done in erecting the wireless mast; it is now practically finished. If the mast itself does not buckle, the stays should hold; but, owing to violence of the prevailing winds, the stays will need overhauling from time to time The wireless operators are busy getting the engine rigged up again.

I then went up the snow slope for about a mile, and, from this point, Winter Quarters looked like a heap of stones. The ice sloped up to the southern sky-line. The dark water to the north was broken by an occasional berg, or a little cluster of ice-covered islands. This wonderful region of ice and sea looks beautiful on a fine evening, but what a terrible vast solitude, constantly swept by icy winds and drift, stretches away to the southward!

I have drafted instructions for the guidance of Mr. Madigan in regard to the despatch of a fully-equipped search party to leave Winter Quarters on the day the "Aurora" sails for the west. The search is to be made in a south-easterly direction, and the extent of the journey will be determined by the leader of the party.

24.1.13. The search party returned from Aladdin's Cave and reported that no traces of the missing members could be found in the vicinity of that depôt.

Supplies and fuel are being landed for the use of the Relief Expedition. All the provisions that can be spared from the stores set apart for the use of the ship's company will be sent ashore during the brief intervals of moderate weather.

A strong gale from the south-east raged from January 25th to 27th. At 9 p.m. on the 25th, the cable parted, sixty fathoms from the anchor. The ship cleared safely the reefs to leeward, and we managed to get in the rest of the chain. By keeping about three miles from the shore in a north-westerly direction, we seemed to be beyond the reach of the more violent gusts.

On the 26th we witnessed a most curious disturbance in progress over the sea to the north. Violent gusts, after approaching our position at great speed, appeared to curve upwards; the water close to the vessel was disturbed, and that was all. After this strange phenomenon had lasted for an hour or so, the wind went into the south-east with a bang. I decided to return to the old anchorage, where we let go our spare anchor with what had been saved of the chain.

On the morning of the 27th, there was a hurricane sky. The chain of the anchor had been severely tested during the night. The wind died away during the lulls, only to blow more fiercely than before, half an hour later. This lasted until 9 a.m., when the wind dropped. The *suddenness* with which changes occur is very remarkable. At 11 a.m. the launch had just left the ship with a cargo of coal when the wind freshened from the south-east. The launch had just got inside the boat-harbour when a terrific gust struck the ship and the chain parted. We were blown out to sea while heaving in the thirty fathoms of chain which remained.

We steamed backwards and forwards until the wind died away, about 4 p.m. Then the launch was able to come off and take a load of stores to the boat-harbour. I decided to steam about for some hours, and return to anchorage early the next morning.

While this gale was in progress, a party of three men had left the Hut to make a prolonged search in a south-easterly direction from Aladdin's Cave. This lasted five days. A record of the journey is given in Mr. Madigan's Report (*see* page 95). This brief record will convey to the general reader some idea of the weather conditions met with, in travelling over the snow-swept plateau.

28.1.13. This morning we let go our kedge weighing about 5 cwt., with the remaining part of the chain. This pulled the "Aurora" up—for a time. A few hours later the gusts caused the anchor to drag, so we were obliged to heave up; when the anchor was in-board, the fluke was missing. . . .

The wireless plant was working last night, but no signals were received. All the gear, coal and food, for the use of the Relief Expedition, is now on shore. I have given them all that could be spared from the ship's stores.

One of the instructions received from Dr. Mawson read as follows :—"Should my party not have reached the Hut before February 1st, you are to steam east, scanning the coast as far as latitude 65° 45′ S. and longitude 145° 50′ E." I shall cruise along the coast to-morrow, if the weather is clear.

LARGE FLOE WITH SEAL. *Photo, Gillies.*

AT "ALADDIN'S CAVE."
A Shelter excavated in the blue ice of the Plateau at an elevation of fifteen hundred feet—the Cave is entered by a vertical shaft situated near the centre of the picture.

MACKELLAR ISLET WITH ICE CAP. *Photo, Gillies.*

THE SECOND ANTARCTIC VOYAGE 93

29.1.13. The weather being favourable we steered to the eastward, keeping from two to three miles from the coast-line. A steep ice-cliff rises vertically from the water until the Mertz Glacier-Tongue is reached. No flag or other sign could be seen on the higher ground. The course followed is shown on the sketch.

Signals were fired at frequent intervals and a large kite was displayed at a height of 500 feet. But all our efforts to attract attention were fruitless.

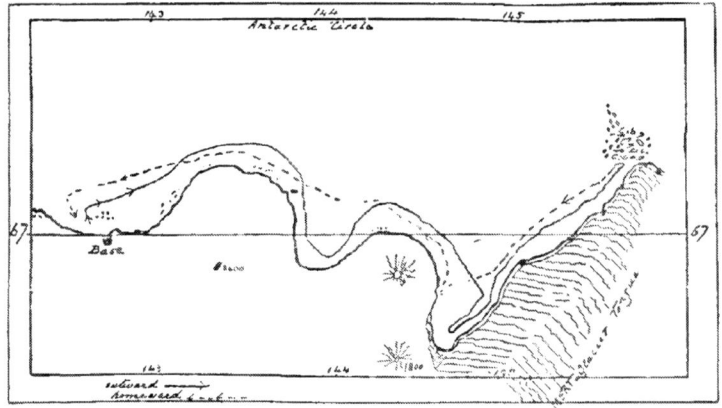

"Scanning the coast."

31.1.13. We returned to the Main Base. I was anxious to sail for the west as soon as possible; the members of the Expedition (not belonging to the Relief Party) were waiting at the Hut, in readiness to embark.

At 9 p.m. we were off the Base. No flag was visible, so we knew that the Leader's party had not come in. They are now sixteen days overdue, and there must be something seriously amiss!

A fresh breeze was blowing from the S.S.E., so we steamed about waiting for the wind to moderate sufficiently to send a boat ashore.

Instead of moderating, the wind steadily increased and, for the next seven days, it blew a continuous, heavy gale, and the temperature fell to 19° Fahrenheit.

The waves rose high, their edges being blown into spume which, as the force of the wind increased, gave place to that grey spindrift resembling smoke, which was the more dense when the velocity of the wind exceeded sixty miles an hour. During the squalls, the velocity was eighty miles, or over. We tried to maintain a position under the cliffs, where the sea was less heavy, and this entailed a constant struggle, as with a full head of steam, the vessel drove seaward, where the waves broke on board, rendering steering more difficult. Then when the wind moderated to a mere howl, we would crawl back; only to be driven out again by the next squall.

Our ordinary speed was six knots, yet for a whole week the ship was being driven at the equivalent of ten knots, without a hitch. We realized what a fine old ship the " Aurora " was, and the efficiency of Gillies and his engine-room staff.

The blinding spray, coming on board, froze solid immediately; until the deck was a skating-rink, and the running gear stiff and hard as iron—each rope being the core of a long icicle. Sometimes when the engines were doing their best —at seventy-five revolutions per minute, the wake of the ship was passing the bow at a rapid rate.

On February 6th, just as the sun was showing over the ice-slopes, the wind became wilder, and the squalls really terrific. The ship was absolutely unmanageable, and driving out to sea. I was expecting the masts to go overboard every minute. This lasted about two hours, when the squalls became less violent. We made shelter gradually, during the morning.

On February 8th, every one was exhausted and the engines were showing signs of the constant strain imposed on them. Suddenly, the wind *ceased*, the weird silence being indescribable, we had become so accustomed to the roar that this silence seemed unreal Would the next squall blow us right out of the water? It came—but with diminished force, and we could rejoice that the gale was moderating. Our struggle was over and victory was in sight—thanks to Providence, a good ship and the stout hearts of her crew.

After 2 a.m. the squalls became much less violent and the lulls of longer duration. At 9 a.m there was only a gentle breeze. We steamed in towards the boat-harbour and

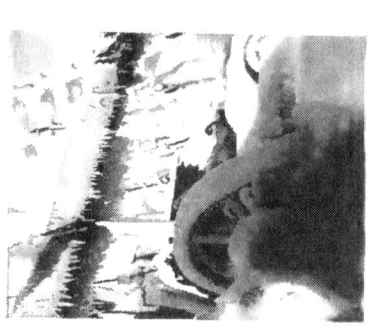

signalled to the waiting members to prepare to embark before noon.

There was no time to lose, in hastening to the relief of Wild's party; in fact it was much later than I considered prudent; but the recent gale had rendered delay unavoidable.

I shall close this chapter with Mr. Madigan's Report on the work of the Search Party (January 25th–29th):—

Winter Quarters,
Commonwealth Bay,
January 31st, 1913.

Captain J. K. Davis,
    Commander, S.Y. "Aurora."

DEAR SIR,—

I beg to report to you that the Relief Party left here in my charge are all well, and we are provisioned for a year with all necessities, and have sufficient clothing and coal. We have the stores sorted and packed, mostly in the store, fuel stacked handy, and sundry improvements made to the Hut.

According to your instructions a party of three men left Winter Quarters on January 25th, at 8.30 a.m., and arrived at the 5½-mile Depôt at 1 p.m., when dense drift compelled them to camp for twenty-four hours. They got away from the Depôt at 2 p.m. on January 26th in moderate drift, and made five miles to the south-east before camping again in heavy drift. They built a five foot mound of snow here, and left provisions and a note giving the bearings and distance of the Depôt. Drift continued all day on the 27th, and travelling was impossible.

They broke camp at 9 a.m. January 28th, in moderate surface drift; it only being possible to see a few hundred yards Course, E. 30° S. After travelling eight miles, a six-foot snow mound was built, with note and provisions as before. Continuing another five miles in the same direction, they camped on the top of a ridge, with a good view. The drift had moderated. An eleven-foot snow mound with note and four days' provisions were left.

The party returned from this camp to the Hut, a total

distance of twenty-three miles, on January 29th, arriving at 5.45 p.m. The 29th was fine, and they got a good view with the glasses. They found no traces of Dr. Mawson's Party. This party consisted of Hodgeman, Hurley and Dr. Mclean.

Two parties are ready to leave Winter Quarters for an extended search when the ship leaves

I beg to thank you for having done all in your power for our comfort for another year in Antarctica, and I can assure you we will be just as well off as last year.

Yours faithfully,
(*Sgd.*) C. T. MADIGAN.

Dr. Mawson writes in *The Home of the Blizzard*, under date January 29th :

" I started on the morning of January 29th in considerable drift and a fairly strong wind. After going five miles I had miraculous good fortune. I was wondering how long the two pounds of food which remained would last, when something dark loomed through the drift. . . . The unexpected had happened—it was a cairn erected by McLean, Hodgeman and Hurley, who had been out searching for us. On the top of the mound was a bag of food—while in a tin was a note stating the bearing and distance of the mound from Aladdin's Cave (E. 30° S., distance 23 miles) . . . the search party had only left this mound at 8 o'clock on the morning of that very day—it was about 2 p.m. when I found it. During the night of the 28th, our camps had been only about 5 miles apart." (P. 269, vol 1.)

Dr. McLean   Bickerton   Bage   Hodgeman   Madigan (leader of Relief Party)
GROUP OF PARTY THAT REMAINED FOR SECOND YEAR.
[Photo Grau.
Page 36.

CHAPTER XII

THE RELIEF OF THE WESTERN PARTY

"Aurora" makes a Start for Wild's Base—Recalled by Wireless. Final Departure for Western Base

> Then take this honey for thy bitterest cup—
> There is no failure save in giving up,
> No real fall so long as one still tries,
> For seeming setbacks make the strong man wise;
> There's no defeat in truth save from within,
> Unless you're beaten there, you're bound to win
> *Marcus Aurelius*

WE left the Bay at 11.30 a.m. and met light airs and smooth sea outside. During the afternoon we steered about northwest. At 8.30 p.m. the ship was approaching heavy pack. The wireless officer appeared on the bridge and handed me the following message * received from the Main Base —

"Mawson returned; Ninnis and Mertz dead; return immediately and pick up all hands."

This message, while naturally a shock, was also something of a relief. The news was definite; and the doubt and uncertainty of the last fortnight were done with.

We turned back towards Winter Quarters, earnestly hoping that the fine weather would continue until we had embarked the party. About midnight a dense mist came on. In such close proximity to the Magnetic Pole the compass is useless, and we had been accustomed to steer by the sun, when visible, or by the direction of the wind; as the coastal wind seldom varies more than two points from S.S.E true.

I could only HOPE that we were steering south; because the ship's head ·by compass was N.N.W., the sun was not visible, and there was no wind.

Shortly after 3 a.m. we could make out the loom of the

* The "Aurora" was fitted with a receiving set, but was without the necessary apparatus for sending wireless messages

ice-cliffs, and the course was then altered. We were some distance west of the point I had steered for; but, under the circumstances, this was a good landfall.

At 8 a.m. on the 9th we were approaching the Main Base, with a fresh breeze, which increased steadily as we drew nearer the land. By noon we were up to the ice-face where the wind was howling and the "Aurora" herself hardly steering. Under these circumstances it was impossible to send a boat ashore; so we signalled for instructions, but could obtain no answer. The Pilot Jack—our pre-arranged signal that Dr. Mawson had returned—could be seen flying from the wireless mast.

We then stood up and down the land, hoping the wind would moderate; and I considered the position in all its bearings.

At 6 p.m. the weather was getting worse, with a falling barometer. I felt that decisive action was necessary. The position was very difficult, as a sense of discipline and obedience to orders urged me to remain, leaving the responsibility on the Leader who had called us back, but DUTY urged me to take prompt action, and I decided to proceed west for the following reasons:—

(1) Dr Mawson and his comrades were safely housed and fully equipped for the coming winter.

(2) Any further delay was seriously endangering our chance of being able to relieve Mr. Wild's party this season. The navigation to the Western Base (1,500 miles distant) was becoming daily more difficult on account of the increasing length of the nights and the conditions of the ice.

(3) The only vessel, "The Gauss," that had wintered in the vicinity of Wild's Base had been frozen in on February 22nd. The "Aurora" was not provisioned for a winter in the ice.

(4) From the records at the Main Base, it had been ascertained that gales often lasted for many days at the close of the short summer season. We had just weathered ONE, lasting seven days.

(5) As a seaman, I realized the difficulties encountered approaching Wild's Base in 1912; and also in getting away from it. It was now THREE WEEKS LATER in the year.

I went down to the ward room and announced my decision to the officers. I invited them to suggest any alterna-

THE RELIEF OF THE WESTERN PARTY

tive measures, but none were forthcoming. At 6.30 p.m. we hoisted the flag and dipped it as a sign that we were leaving. Sincere regret for the detention of our Leader and his six comrades was combined with grave anxiety as to being able to reach the Western Base in time. Until Wild and his party were on board, it was impossible to tell whether we should not miss both parties. Few more difficult situations have arisen in Polar work, but I was convinced that, in leaving the Main Base, without further delay, I was acting as Dr. Mawson would have wished, had he known the position of the Western Party.

We steamed out of the Bay, and the wind moderated as the ship got well out to sea. At midnight there was a moderate breeze from the south, with some snow.

10.2.13. About fifty miles north of Commonwealth Bay we met some heavy pack. After skirting this pack for a time, we pushed into it, and reached open water at 1 p.m., after three hours amongst the floes.

At 2 p.m. we sighted what looked like a large mass of barrier-ice adrift. Was this a portion of the barrier sighted in January, 1912 ? We had noticed the disappearance of this "barrier" (from the position occupied in 1912) while approaching the Main Base on January 12th, 1913. It was in a position about twenty-seven miles to the north-west, and the face, along which we sailed to-day, was quite forty miles long.

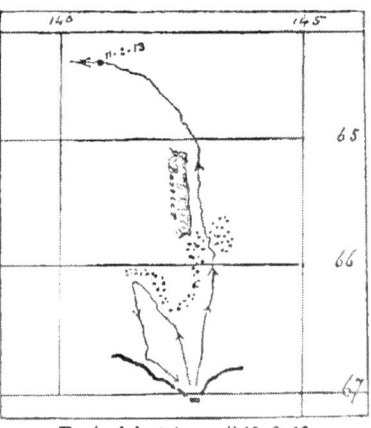

Track of the "Aurora," 10.2.13.

The wireless officer picked up a message from Dr. Mawson to-night, but only a few words could be made out ("crevasse," "Ninnis," "Mertz" ? "broken," "cable").

11.2.13. We are evidently north of the main coastal pack, but a high westerly sea makes progress slow.

13.2.13. Three days of westerly winds and high sea. At noon the pack was visible to the southward. We are making little more than sixty miles a day towards our goal!

14.2.13. This afternoon an easterly breeze sprang up; all sail was set immediately, and we were soon making about seven knots. The record for the last three days has not been encouraging.

15.2.13. We are running before an easterly gale, in thick snow. We are making a fair course to the west using the same errors as were recorded last voyage. At noon to-day we were 180 miles to the good. The "Aurora" is now doing eight knots, but the driving snow makes it difficult to see anything in the way of obstacles.

This gale continued for three days, with very brief intervals of clear weather; and good progress was made towards Termination Tongue. Prudence urged a reduction of speed. Should any obstacle appear suddenly under our bow, a collision, in the sea that was running, would mean destruction. I must acknowledge taking heavy risks on this occasion; but the circumstances rendered such imperative; when a fair wind gave us an opportunity of a good run and making up for lost time, it had to be embraced—*at all costs.*

To be successful in getting through the ice to the west of Termination Ice-Tongue we must reach that locality before new ice which will consolidate the pack, has formed. At night, the difficulties of navigation are increased—you are unable to see whither you are trending, and daylight may reveal the fact that your vessel is fast for the next twelve months! The "Gauss" was BESET by ice on February 22nd.

18.2.13. To-day the weather cleared and we were able to get good observations. Our position at noon was in latitude 64° 25′ S., longitude 105° 14′ E. Several bergs were in sight, suggesting heavy pack to the southward. A great many blue whales have been seen during the forenoon. Petrels and Cape pigeons are flying about, in scores.

19.2.13. We were brought up this morning by a line of dense pack across our course. The weather was misty, but cleared up before noon. We had to follow the pack-edge in a north-westerly direction until daylight on the 20th.

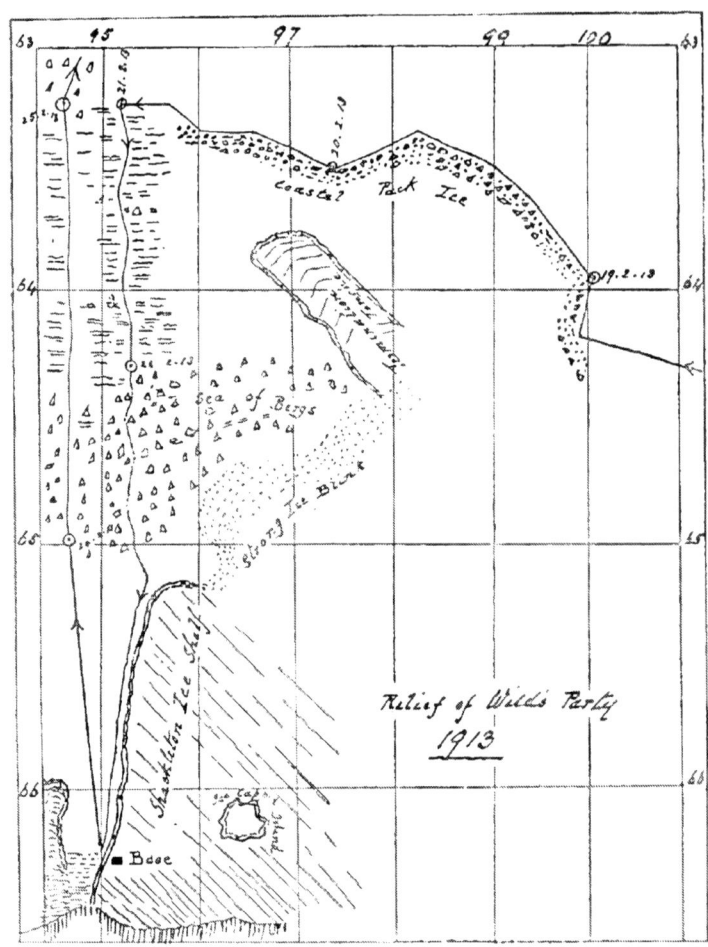

The relief of Wild's Party, 1913.

This heavy pack is some sixty miles farther north than the edge followed in 1912.

20.2.13. At daylight we were able to steer south-west. At noon we were about twenty miles north of Termination Ice-Tongue. We tried to push through the pack but it proved impenetrable, so we continued farther west where the sky looked more promising. There was some snow, but the wind continued to blow lightly from the east.

21.2.13. At noon we pushed into the looser pack and made good progress to the south   At 8 p.m. we were steering through leads by moonlight.

22.2.13. At 4 a.m. the wind freshened from the south-east with some snow. The floes were getting heavier and progress was slow. After having worked through some eighty miles of heavy floes, which were, most fortunately, separated by leads of open water, we reached the confines of that berg-laden sea, where the navigation had proved so dangerous in 1912. The bergs were now so numerous, that there was some difficulty in avoiding a collision, even during daylight. . . .

In the afternoon, a blizzard came on, and at 8 p.m. the darkness and the falling snow made it impossible to see any distance ahead. . . .

For the next seven hours we threaded a passage through this sea of bergs without mishap, guided and protected by a Higher Power. . . . Never was dawn awaited with more anxiety, nor its appearance hailed with more thankfulness. . . .

23.2.13. At 4.30 a.m. the loom of the Shackleton Shelf was sighted; the falling snow ceased, and a very strong ice-blink was observed to the eastward, and to the north of the Shelf. We stood in to follow the ice-cliffs towards the Western Base.

At 11 a.m. we sighted a depôt flag near the Base. As we were approaching the edge of the sea-ice at the foot of the ice-cliff we were able to count eight figures on the floe; suddenly, these figures hastened to the edge and dived into the water. Then we realized that this " party " consisted of Emperor penguins!

As we drew closer, we caught sight of Wild and his comrades. Some one on the forecastle head shouted " Are

(Back row)   Jones   Moyes   Hoadley   Kennedy

(Front Row)   Harrison   Watson   Dovers   Wild
[Photo, Gillies.
THE WESTERN PARTY TAKEN JUST AFTER RELIEF.

[Photo, Hurley.
PHOTO OF "ALL HANDS" TAKEN AFTER THE RELIEF OF WILD.

THE RELIEF OF THE WESTERN PARTY

you all well ? " and it was good to hear Wild's cheery " Yes, all well, and glad to see you." Then came the question : " Is everything right at the Main Base ? " Our silence told of bad news. In a few minutes the party had clambered on board and question followed question. It was with a feeling of profound relief that I shook hands with Wild, and realized that, although we had left them in a perilous position, we had, in spite of all difficulties, been able to welcome them heartily on board the " Aurora " to-day. I think that I understand the feelings which prompted the old Arctic Explorer who named " Thank God Harbour."

Wild and his companions were very glad to see the " Aurora." They had commenced to lay in a stock of seal meat, against the event of having to spend another winter on the barrier. We were very busy all the afternoon embarking specimens, stores and baggage, and replenishing our water tanks. Our recent experience had shown clearly that we could not afford an hour's delay before recrossing the berg-laden sea and the long line of pack which lay to the north of it. These were two serious obstacles lying between our present position and the open sea.

I decided to spend the night steering north under the shelter of the barrier, so as to recross the Sea of Bergs in daylight ; and, if possible, before the advent of another blizzard. We were all truly thankful that Wild and his comrades were safe on board ship and every member of the party in good health.

We sailed north at 9 p.m., taking advantage of a fair moon to lighten our darkness.

The inset overleaf shows the track of the " Aurora " from February 19th to February 26th, in the vicinity of Termination Ice-Tongue and Shackleton Ice-Shelf.

To the south of the 65th parallel open water was found.

24.2.13. We are now crossing the Sea of Bergs, but, this time, in fine clear weather. The bergs are composed, for the most part, of glacier-ice.

It is simply marvellous how a vessel can pass through such an accumulation, IN THE DARK, and escape with only a few bumps. At 4 p.m. we entered the pack of heavy floes, the leads were fairly open and the pack loose.

Wild's party had experienced a great deal of bad weather, which had hampered their movements on shore, but they succeeded in doing some excellent work. This party, whose landing and relief afforded me opportunities of knowing them well, I shall always regard with admiration for the plucky way in which they agreed to land on floating ice seventeen miles from the coast-line rather than return to Hobart, leaving their work undone. The mystery of the high land sighted by Drygalski has been cleared up—it has proved to be an ice-capped island. The party discovered a large glacier about 120 miles east of their Hut. This icy stream, falling three thousand feet in fifteen miles, and

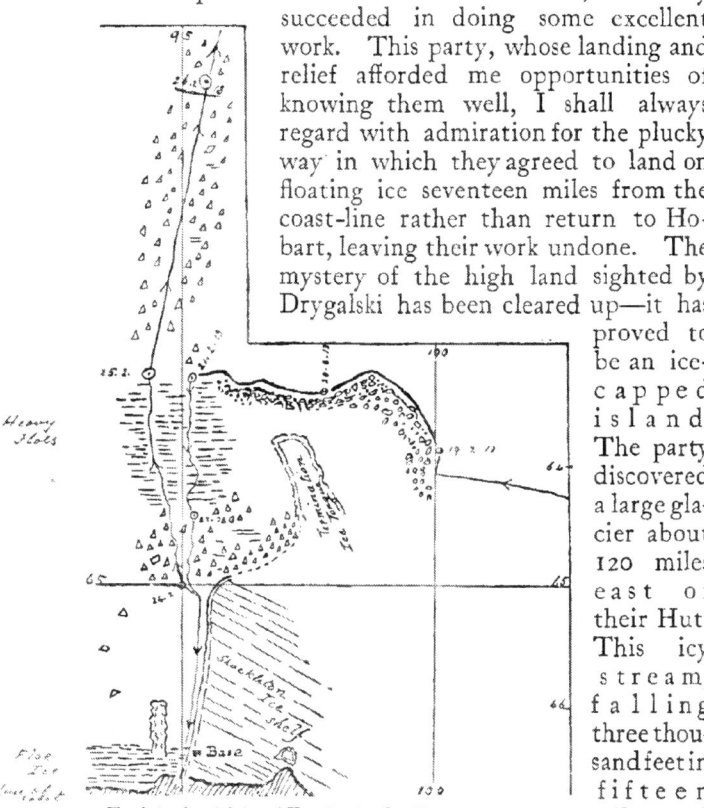
Track in the vicinity of Termination Ice-Tongue.

pushing out through the Shackleton Shelf, "sweeps on to the northern horizon in ever-widening billows of upturned impassable ice."

Three members of the party had journeyed to the east and reached Gaussberg; there connecting with the work of the "Gauss" Expedition. For details of their work I must refer my readers to the "Home of the Blizzard." The winter had been severe, and the Hut was covered with drift-snow. On fine afternoons all hands would visit the floe and take some exercise on skis; this was the favourite recrea-

A HEAVY FLOE [Photo, Gillies.

PENGUINS ON THE SEA ICE OFF WILD'S BASE. [Photo, Davis.

tion, and one provocative of much mirth if the "slope runner" rolled over and over in clouds of soft snow.

25.2.13. At 8 a.m. we had cleared the pack, and kept on a northerly course through a collection of bergs, many of which were earth-stained. Fine weather and light winds. . . .

26.2.13. We are still steaming through a sea with numerous bergs, of worn appearance. We have now come through over three hundred miles of berg-laden sea, and this morning they are still visible all round. We have been most fortunate in getting fine weather since leaving the Western Base

There appears to be some current or meeting of currents which causes the ice to collect in this locality. I have seen nothing like it in any other part of the Antarctic

The last ice-bergs were sighted in latitude 55° 30′ S.

Fresh westerly gales and high seas were experienced during the first week of March. Hobart was reached on March 15th. We heard that wireless messages had been received from Dr. Mawson. . . .

We also learned, with deep regret, that the "Terra Nova" had reached New Zealand in February bringing the news of the sad disaster which had overtaken Captain Scott's party on March 25th, 1912. Successful—defeated only by death—Captain Scott and his companions sleep on the field of their achievements.

Dr. Nansen well expressed a universal thought when he said of these men, "Their shields are as bright and shining as the snow that covers their graves."

CHAPTER XIII
THE MAWSON RELIEF FUND, 1913.

As soon as possible after reaching Hobart, I reported to Professor David, who had been in communication with Dr. Mawson.

A wireless message from Dr. Mawson had reached Sydney on March 5th, 1913, in which the main features of the disaster are described :—

" On December 4th (1912), while exploring new coast-line three hundred miles south-east of Winter Quarters, Lieut. Ninnis, with a dog team and almost all our food, disappeared in an unfathomable crevasse.

" Dr. Mertz and myself started, over the plateau, for our Hut with inadequate supplies and six starving dogs

" On January 17th (1913), Dr. Mertz died from causes arising from malnutrition.

" On February 7th, I alone reached the Winter Quarters, having travelled through snow and fogs over heavily crevassed areas, miraculously guided by Providence.

" The ' Aurora ' had waited until it was no longer safe, and left only a few hours before I had reached the Hut. Six men were left here to prosecute a search."

The unavoidable extension of the Expedition for another year would involve additional expenditure. The Committee, acting as Dr Mawson's representatives (in Australia), decided that the " Aurora " should be laid up during the winter of 1913, and that I should proceed to England with the object of :—

(1) Reporting progress to our supporters in that country, and—

(2) Raising additional funds for the relief of the party in Antarctica.

THE MAWSON RELIEF FUND, 1913

I reached England in May, not without some misgivings about being able to carry out the wishes of the Committee with regard to raising funds, but determined to do my best. Candidly, I should have given up the campaign, after the first two weeks, had it not been for the kind assistance and encouragement of Dr. H. R. Mill, who has done so much to help forward the cause of Antarctic exploration. His keen enthusiasm and kindly sympathy—as he talked of the prolonged struggles which former Explorers had waged with apathy and indifference, before the campaign was won— would have cheered the most despondent, and inspired the most faint-hearted collector to " *Carry on.*"

Nevertheless, it was uphill work; those who understood the circumstances and were sympathetic had no money. I was getting one rebuff after another. when Sir R. Lucas-Tooth came forward and started the Fund with a cheque for £1,000. It was not only the handsome donation, but also his kindly advice and assistance that gave me further encouragement to persevere.

I issued an appeal inviting contributions to " The Mawson Relief Fund, 1913," addressed to " all who sympathize with organized and determined effort in overcoming difficulty, especially the difficulties incident to exploration."

At a later date, I was received by the Right Honourable David Lloyd-George (then Chancellor of the Exchequer), who promised that the matter of a grant to the Relief Fund would be considered when framing the estimates. Before I left England I was informed by the High Commissioner for Australia (Sir George Reid) that the Imperial Government had decided to grant £1,000 to the Fund.

The late Sir John Murray took a kindly interest in the Relief Expedition, and his outspoken appreciation of what had been accomplished by the Australasian Antarctic Expedition was of great value.

Lady Scott, to whom I had been charged to convey a message of condolence from our Committee, not only took a keen interest in the appeal, but insisted on contributing liberally herself, as did Commander Evans and the members of the Scott Expedition.

I left London early in August, and on reaching Melbourne reported to the Committee, who then arranged for a deputation to wait upon the Prime Minister of the Commonwealth, Mr. J. Cook (now Sir Joseph Cook, P.C., G.C.M.G.), and place the position before him. Mr. Cook remarked to the deputation that he considered "Bringing back Dr. Mawson was an Australian responsibility." A grant of £5,000 from the Commonwealth Government was available for the purpose, within forty-eight hours. This, in addition to the contributions from British supporters, solved the financial problem.

To those kind friends in London who came to the assistance of the Australasian Antarctic Expedition, I can only repeat the words of Hamlet.—

"Beggar that I am, I am even poor in thanks"

The "Aurora" was brought to the docks at Williamstown, where, for the second time, the Victorian Government acted in the most liberal manner with respect to refitting. A new tail-shaft was fitted and various repairs executed. Prior to our departure from Williamstown, we were honoured by a visit from the Governor-General, Lord Denman, who inspected the ship and wished us God-speed

The "Aurora" sailed from Hobart on her third voyage to the Antarctic on November 19th, 1913.

[Photo, Davis.
THE "AURORA" IN DOCK SHOWING THE HEAVY FOUR-BLADED PROPELLER.

THE "AURORA" IN DOCK AT WILLIAMSTOWN, VICTORIA. [Photo, Murphy.

CHAPTER XIV

THE THIRD ANTARCTIC VOYAGE,

NOVEMBER 1913 – FEBRUARY 1914.

"Yet all experience is an arch wherethro'
Gleams that untravell'd world, whose margin fades
For ever and for ever when I move"
Tennyson.

19.11.13. The "Aurora" left Hobart to-day on her third voyage to the Antarctic. In addition to the ship's company, Messrs. Hurley and Correll, photographers, and Mr Hunter, biologist (all of whom returned from Adelie Land last voyage), are on board. The Commonwealth Government has arranged to take over and maintain the wireless station established at Macquarie Island two years ago, and Messrs. Power, Henderson and Ferguson are coming with us to relieve Messrs Ainsworth, Blake, Sandell and Hamilton, who have been at the station since December, 1911. We are taking seventeen sheep to be landed at Macquarie Island, where they will soon fatten on the excellent pasture.

After landing the relief party, we propose to spend some days in the vicinity, taking soundings to supplement the information embodied in the excellent map of the island recently made by Mr. Blake. As too early an arrival at Commonwealth Bay would only mean getting into the pack, thus causing delay, I consider that some days spent in sounding off the island, if weather permits, will not be time wasted.

21.11.13. Since leaving Hobart we have been favoured with remarkably fine weather. We are taking a southerly course, so as to pass over the rise discovered during the spring cruise of 1912. At 4 p.m. to-day we sounded in 640 fathoms on hard bottom. After recording the depth we lowered a rock-gripper, but failed to obtain a specimen.

22.11.13. At noon we sounded in 2,400 fathoms on mud; an important sounding which apparently defined the southern limit of the rise. Large quantities of kelp have been passed during the day. Large masses of this substance, which have probably drifted from Kerguelen Island (3,130 miles to the westward), are frequently met with between latitude 49° and 52° S.

23.11.13. A fine day with light airs and very little swell. Dull, cumulus clouds all round on the horizon. At 1 p.m. we sounded in 2,470 fathoms on ooze. We then took a number of temperatures and samples of sea water at various depths.

Latitude 50° 30′ S
Longitude 148° 2′ E

November 23rd, 1913.

SERIAL TEMPERATURES

Taken with Ekman reversing bottle fitted with Richter Thermometer No 672.

Fathoms	Centigrade	Fahr
Surface.	8.48	47 26
10	8.20	
25	8.25	
50	8 19	46 74
100	8 23	
150	8 34	
200	8 29	46 92

* "Ocean currents are progressive movements of the water, due partly to prevailing winds, and partly to difference of temperature and density of the waters. These ocean streams maintain a constant circulation and interchange, the movements taking place in a horizontal as well as in a vertical direction." The study of this ceaseless circulation in oceanic waters is important, because the ocean radiates its heat more slowly than the land, and, its area being three times greater, it becomes a vast storehouse of heat, which the waves, tides and currents distribute over the surface of the globe. The combined elements that affect the weather of the habitable land are included in the term "climate." Of these elements, temperature and moisture are probably the most important.

The Gulf Stream leaves the Gulf of Mexico as a river of very salt water, fifty miles wide and three hundred and fifty fathoms deep, having a surface temperature of 81° F. (27.2° C.).

* "Wrinkles in Practical Navigation," S T S Leckey   Geo Philip & Son

It flows along the coast of Florida at five miles an hour, but off Cape Hatteras curves towards the east as far as the Azores, whence the greater volume flows to the north-east, passing the British Isles, and bringing warmth and moisture to North-Western Europe. It is in a large measure due to the influence of this great current that the climate of Britain is so mild compared with other regions in the same latitude.

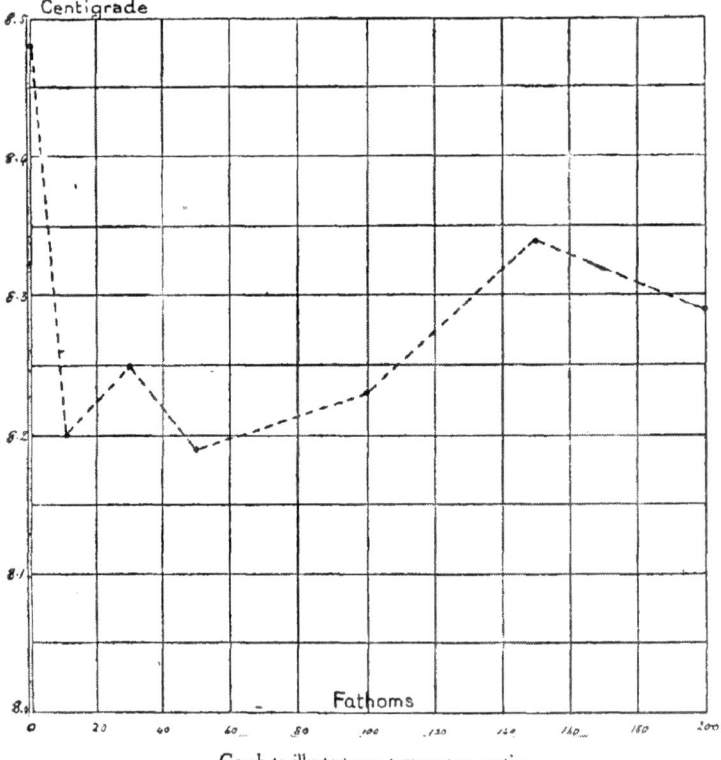

Graph to illustrate sea temperature section.

It is the business of the oceanographer to map these warm or cold streams on the charts of the oceans. Data are now being collected with a view to extending a knowledge of the subject. In our case we were working over a new area, where every depth-and-temperature section would be

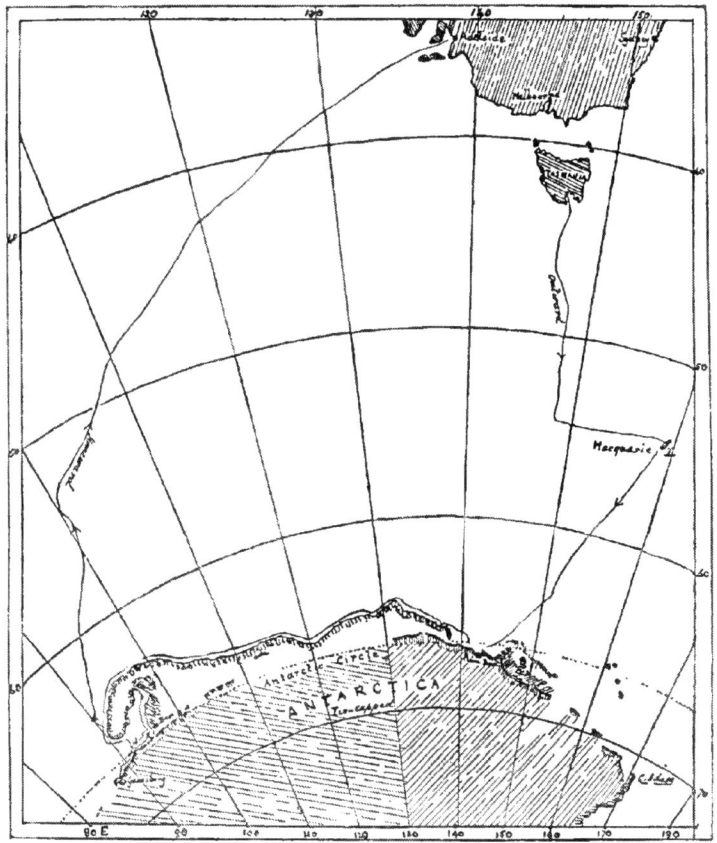

Relief Voyage Hobart to Adelaide, 1913–14.

The sketch shows the track followed by the "Aurora" on the third Antarctic voyage from Hobart. The dark line shows the position of *terra firma* between Cape Adare and Gaussberg.

The wide belt of ice fringing the continent in the season 1913–1914 prevented the ship from approaching the coast-line in the vicinity of "Knox Land," reported by Wilkes in 1840. From the information obtained by Wild's party there is every reason to assume that the actual *coast-line* is continued in an easterly direction.

valuable, as coming from a part of the ocean where nothing of the kind had been attempted hitherto.

The party who carried out the deep-sea work under my direction was headed by the Chief Officer, Mr. Fletcher, and all were deeply interested in bringing it to a successful issue. Taking soundings in high latitudes is not a pleasant job, but with unanimity and determination, good results may be obtained. I shall always remember the "sounding party" on the "Aurora" as a small band of cheerful and energetic enthusiasts, ever ready for action, and reliable under the most trying conditions.

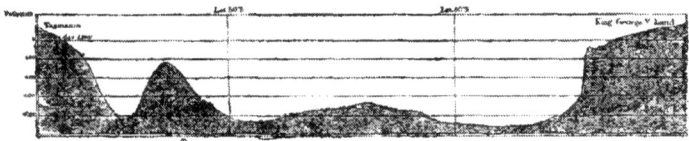

Section of the SEA FLOOR between Tasmania and the Antarctic about the 149th Meridian from soundings taken on board the "Aurora." Australasian Antarctic Expedition, 1911–14.

24.11.13. Latitude 52° 18′ S., longitude 148° 10′ E.

We are now south of the zone, between Tasmania and latitude 52° S., where, on previous occasions, we have always met with the worst weather of the outward voyage. So far we have not had a single gale. To-day the weather conditions are ideal; a light south-east wind and a smooth sea. The fore hatch has been opened, and we have shifted some coal into the bunkers. Early this morning several penguins were observed in the water close to the ship.

25.11.13. Fine weather with a light easterly wind. During the forenoon we sounded in 2,300 fathoms, and then took a series of water-temperatures and samples. The surface temperature showed a fall of 6° since our last series taken on November 23rd. At noon Macquarie Island bore approximately east, 380 miles distant. A course was then set to approach the island on this bearing. It was possible that traces of another submerged land mass might be found west of the Island.*

* Referring to the geological examination of Macquarie Island by Blake, it is stated that the "Island has been overridden, comparatively recently, by an ice-sheet travelling from the west" (*The Geographical Journal*, September 1914, p. 283).

26.11.13. Moderate breeze. Cloudy weather. Barometer 29.18. At noon we sounded in 2,220 fathoms on a hard bottom. The driver was recovered, but no sample was brought up. A series of temperatures was taken. At the surface it registered 3.39° C., whereas, at one hundred fathoms, it was 3.33° C.

27.11.13. Fresh westerly breeze with light snow-squalls. At noon Macquarie Island bore east, 162 miles distant. A sounding was made in 2,340 fathoms on hard bottom. The driver was recovered, but no samples obtained.

In a number of soundings taken west and south-south-west of Macquarie, Island we were unable to bring up any sample in the driver. In such cases the bottom was recorded as "hard." Previously, in depths of over 2,000 fathoms, a sample of ooze had, in most cases, been recovered. This absence of sample in the vicinity of Macquarie Island appears to me rather strange. Is the sea floor covered with a deposit too hard for the driver to penetrate?

In 1911 the "Terra Nova" reported a depth of 3,000 fathoms on rock. It would be interesting to know whether a specimen of the rock was recovered on this occasion. The sounding was taken in the same latitude, approximately, where we obtained a depth of 2,000 fathoms on hard bottom, ninety miles further west.

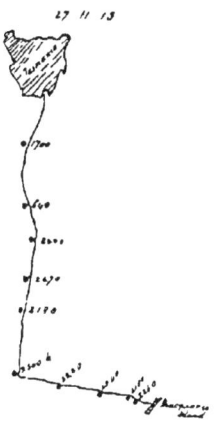

Hobart to Macquarie Island

28.11.13. At 7 a.m. we sounded in 2,180 fathoms. A small black stone came up in the driver, but there were no signs of ooze. A second sounding at noon gave 2,269 fathoms, but no sample was recovered.

At noon Macquarie Island was about thirty-six miles to the eastward. A curious cumulus cloud overhanging the island attracted attention. This cloud formation was invariably looked out for by sealers when approaching the land, which was seldom visible from a greater distance than eight miles, except when south-westerly winds had cleared the misty atmosphere. To-day there is a fresh southerly breeze, and the land was visible from a distance of nearly thirteen miles.

We steered for Hasselborough Bay and anchored in twelve fathoms. As the wind died away, a mist crept over the hills. I decided not to send a boat ashore to-night, as all hands will be wanted at 6 a m to-morrow. A message was signalled to the shore party by Morse lamp, and this had the anticipated result of bringing our comrades off in their own boat, which they dragged across the isthmus from North-East Bay. They reported all well at the Shack, and gave us the latest items of wireless news. We heard that the " Rachael Cohen "—the sealers' vessel—had sailed to-day for Hobart with a cargo of oil.

CHAPTER XV

THE LAST VISIT TO MACQUARIE ISLAND

"A wanderer is man from his birth.
He was born in a ship on the breast of the River of Time,
Brimming with wonder and joy"
                                            Matthew Arnold.

29.11.13. At 6 a.m. we commenced landing the stores and equipment which we had brought down for the use of the Commonwealth Party. By noon we had landed the larger portion, when the surf became so bad that we had to hoist the boat up and await improvement. The swell is very confused, and the surf is heavy on both sides of the island.

I had a long talk with Mr. Ainsworth, who has collected a vast amount of meteorological data during the last two years. He mentioned that the maximum temperature recorded during that period was $51°.8$ F.$=11°$ C.; and the minimum $26°$ F.$=-3°.33$ C. Just now I am interested in getting a forecast of the probable direction of the wind in the immediate future, as during November and December it is extremely variable in direction as well as in force. At all times of the year one must be prepared to get under weigh at once, if a sudden change of wind finds the ship anchored off a lee shore.

As a result of Blake's work, we are now provided with a good map of the island. Our soundings plotted thereon give a good general idea of the depths in the vicinity of each anchorage. This is a great improvement since my first visit in the "Nimrod" in 1909, when the only chart available was a sketch—hopelessly incorrect. To the seaman provided with a good chart, the difficulties of navigation in these latitudes are reduced to a minimum. As a practical illustration of the value of Blake's chart, I may mention that when two lights were shown from the higher ground of the isthmus, I was able to enter the North-East Bay after dark (June 20th, 1912), during a

THE LAST VISIT TO MACQUARIE ISLAND

westerly gale, and bring up without difficulty by cross-bearings of the lights Any such attempt based on the old sketch chart would have probably ended by bringing up *on the rocks*. What I have just written must not be taken as detracting from the work of the pioneers who first charted the island. Their business was to get seals, while ours is to provide a correct representation of the natural features of the island for the guidance of future visitors.

From observations taken 30.11.13. the position of our anchorage is Lat. 54° 30′ 36″ S. Long. 158° 58′ 0″ E. Anchor Rock bearing N.E. Magnetic,* 4 cables.

This position I can recommend to any ship visiting the island in November or December and finding the wind Easterly. The holding ground is good, but as bottom is uneven and rocky, a good length of chain is necessary

2.12.13. Yesterday a fresh breeze and a confused swell setting into the bay made boat work impracticable. At 6 a.m. to-day the weather was fine and the surf had moderated considerably. During the day the remaining equipment of the Commonwealth Party was landed, and by 9 p.m. we had embarked the specimens and collections of the A. A. E. Party. The weather has been perfect throughout the day, and I was able to send a Wireless message to Mr. Hunt, the Commonwealth Meteorologist :—

"All stores for your party safely landed. Excellent passage from Hobart. Your forecast correct."

I hope to commence sounding to-morrow.

3.12.13. A dense mist obscured the land until 9.30 a.m., when it cleared sufficiently to get under weigh. We then ran a line of soundings out of the Bay and down the western side of the island at intervals of one hour. Former sounding work had been done mainly in the vicinity of two northern anchorages and in the passage round North Head. On the western coast the chart was almost a blank as far as depths were concerned. In fact, this coast had been usually avoided as a dangerous lee-shore.

When we anchored in Lusitania Bay at 9 30 p m. ten additional soundings had been recorded. On looking over the chart, one could not but "Contrast the petty done"

---

* Magnetic variation 18° 23′ E. (Macquarie Island, 1913)

with "the undone vast"; a reflection which always helped us to avoid becoming too pleased with ourselves.

The section shows the depths at certain points on a line running E. and W. through the island, half a mile north of Prion Lake. The sketch is approximate; and, on the line indicated it does not pretend to show the depths at intermediate points. From the western coast line, at a distance of 5 miles, the depth is 140 fathoms. At a distance of 35 miles the depth is 2,200 fathoms. From the eastern coast, at a distance of 5 miles the depth is 1,548 fathoms. At a distance of 10½ miles, 2,735 fathoms of wire were paid out without finding bottom.

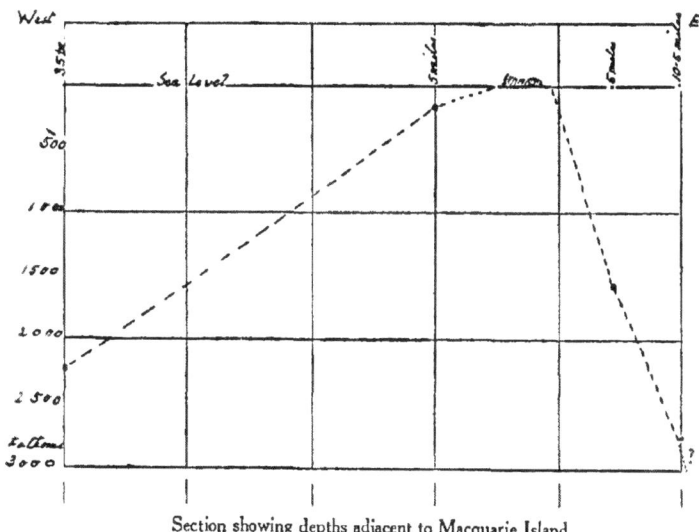

Section showing depths adjacent to Macquarie Island.

4.12.13. At 8 a.m. there was a heavy easterly swell, but the wind remained N.N.W. A dense mist hung over the land until noon, when the wind freshened up and cleared it away. It also smoothed the swell, so I decided to make an attempt to get some cases and casks of specimens, which were ready for shipment, on board during the afternoon. The boat was put over and we pulled in to have a closer view of the surf. It certainly had an ugly look as it broke well out, but,

THE LAST VISIT TO MACQUARIE ISLAND

as the boat was an excellent one, we succeeded in running through it to the beach.

On the first trip we took all the cases and towed three casks out with a long line. On the second trip we made a raft of the nine remaining casks, which were rolled to the water's edge. A long line, fastened to the first cask, was then passed out to the boat, which was pulled outside the surf. After waiting a short time, a heavy roller floated the raft. Then came a tug-of-war. It was some time before we could be sure the raft was not going to pull the boat back into the surf, but in the end, by dint of hard pulling, the raft was towed clear and brought alongside the "Aurora." Everyone was wet through, but we were all very glad to have got all packages on board without damage of any kind.

As a landing place Lusitania Bay has a bad reputation. It is about the worst on the island.

5.12.13. We left Lusitania Bay at 6 a.m. and steamed up the east coast against a fresh N.N.E. wind, with heavy rain. The land was hidden by mist until 11 a.m. when the wind shifted to the north-west and the weather cleared.

We resumed sounding operations off Brothers Point and continued them until we anchored in N.E. Bay. We pulled ashore and landed without difficulty. The Commonwealth Party had taken possession of the Shack, which had been renovated to suit the tastes of the new occupants. To-day it looked bright and comfortable, although the weather outside was cheerless.

I climbed up North Head, whence there is a clear view in moderate weather, to see if I could detect any signs of broken water on the line of the submerged reef. Although there was a heavy swell I was unable to see any broken water between the northern point of the island and the Judge and Clerk rocks.

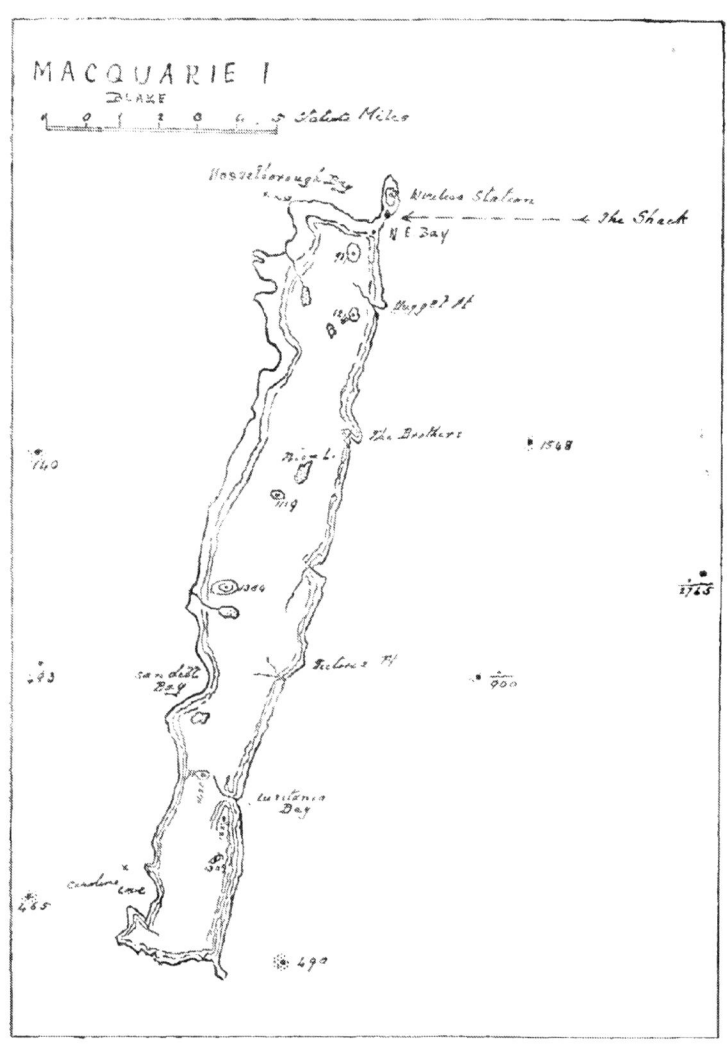

Blake's chart of Macquarie Island.
(Reduced scale).

CHAPTER XVI

FROM MACQUARIE ISLAND TO COMMON-WEALTH BAY

"For we're booming down on the old trail, our own trail, the out trail,
We're sagging south on the Long Trail, the trail that is always new"
Kipling

6.12.13. At 10.30 a.m. our position was in latitude 55° 43½′ S., longitude 157° 59′ E. on a course S. 28 W., when we sounded in 2,420 fathoms on a hard bottom. The driver was recovered, but no sample obtained. The ship was making seven and a half knots under square sail at noon, with a strong breeze from the N.N.W., accompanied by rain squalls.

Mr. Ainsworth has kindly volunteered to keep a record of cloud observations. As he is an experienced meteorologist, his notes will be a valuable addition to those of our Second Officer, Gray, who takes the regular observations.

7.12.13. Our position at noon was in latitude 58° 6′ S., longitude 155° 55′ E. This made our distance from the Main Base about 646 miles. The wind is from the north-east and moderate in force. Blue Billies* and Mother Carey chickens were flying about the ship during forenoon.

We are now making good progress on a direct course to Commonwealth Bay. At 4 p.m we sounded in 2,000 fathoms on a hard bottom A small pebble was recovered in the driver, but no mud. The "Terra Nova" recorded a depth of 3,000 fathoms about ninety miles further east, so the bottom appears to be very uneven ; perhaps there is land in the neighbourhood. So far as I can ascertain, we are steering over a new track Great numbers of Blue Billies have been sighted yesterday and to-day. It would be a piece of luck to discover an island about here ; indeed one great fascination of a voyage like

* Prion Banksii.

this is the possibility of stumbling on some feature hitherto unknown.

8.12.13. At noon we were 547 miles from the Main Base in Adélie Land. The wind was light northerly until 3 p.m., when it freshened to a moderate gale from the north-east. The sky is dull and overcast, while continuous light rain makes everything cold and cheerless. However, the ship is bowling along under topsails and foresail this evening.

Our sounding to-day showed a marked rise—1,560 fathoms on a hard bottom. The surface temperature of the sea water is 2.32° C. Everything works smoothly on board ; the officers know their work and the men are willing. It seems a pity that just as all hands are efficient and well trained for the work of an exploring vessel, the Expedition comes to an end. However, we have this season still ahead, and we must make the most of our opportunities.

9.12.13. After a fresh gale during the night the wind shifted to north by east this forenoon. Thick mist came on about 2 p.m. which obliged us to take in foresail and keep a very sharp lookout. At 9 p.m. we passed several small pieces of drift-ice. Mist continued very thick and, as a precautionary measure, the lower topsail was taken in at 8 p.m. A berg would only be visible at a *very* short distance, so we are just making about five knots under bare poles ; but it is better to be sure than sorry in icy latitudes.

10.12.13. We sighted the first bergs at 5.30 a.m. in latitude 63° 33′ S., longitude 150° 29′ E. At 6 a.m. we sounded in 2,100 fathoms The weather was overcast, with some snow. At noon we were 262 miles from the Main Base. We altered course to S. 32 W., hoping to find some loose pack off the end of the Mertz Glacier-Tongue, through which we can push into the comparatively open coastal water.

The weather continues thick with moderate northerly wind and occasional snow showers. I think this northerly wind is fairly constant just north of the pack. Last voyage we met a heavy northerly swell on the edge of the pack on several occasions. This, no doubt, helps to keep the ice south in spite of the coastal hurricanes. Between 8 and 10 p.m. we passed several water-worn pieces of ice, but very few bergs The pack is probably not far distant.

10.12.13. At 2 a.m. we encountered heavy pack and

PREPARING THE WHALER.

THE BIOLOGIST

Photo, Gillies.

bergs. The weather being too thick to determine its extent, we hauled out to the westward and reduced speed. We were able to make about south-west through thick drift ice all the forenoon. At 4 p.m. the weather cleared and we stood to south-west at full speed. At 11 p.m. there was a strong blink to the south-east. Numbers of snow and Antarctic petrels flew about the ship. A gentle breeze and clear weather made the surrounding ice look very beautiful in the different lights. At midnight the engines were reduced to half-speed as the ice looked closer to the southward. The compass has become sluggish again.

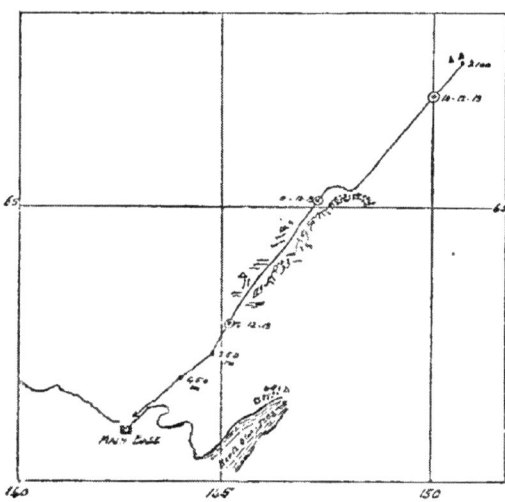

Approaching the Main Base, 1913.

12.12.13. We were able to keep moving all last night, the pack becoming looser as we advanced southward. At noon we were in fairly open water with occasional long streams of drift-ice and a few large bergs. During the whole passage through the pack we have had a good water sky to the southward—probably owing to the fact that there is so much open water between the floes—which is remarkable at this early part of the season. We observed a very strong blink to the south-east and a similar appearance was noted in the same locality on our first voyage. At 2 p.m. the sky cleared and land was sighted to the southward; the well-remembered snow slopes gleaming in the sunlight. At 8 p.m. we were forty-five miles from the Main Base, when we sounded in 450 fathoms on mud. At midnight we were off the eastern point of Commonwealth Bay with a light S.S.E. breeze and fine clear weather We hope Dr. Mawson and his comrades will breakfast with us to-morrow morning.

13.12.13. 3 a m. We are approaching our destination and can distinguish the Mackellar Islands. The weather continues fine, but the snow slopes look bleak and dreary. It is just eleven months, almost to the hour, since our arrival last year. We are a month earlier this time and have had very little trouble with the pack.

At 7 a.m we anchored in Commonwealth Bay in twenty-five fathoms. High up on a rocky promontory we observed a large cross erected to the memory of Mertz and Ninnis. There it will remain as a witness that the silence of the vast white solitude, undisturbed since Time began, has been broken, and of the price paid by the pioneers.

The whaler was lowered and we pulled ashore. We rounded the point of the boat harbour in silence, and moved quickly towards the Hut. Mawson appeared in the doorway, and on seeing us gave a loud shout, " Halloa, you chaps, the boat is here." The others tumbled out of the Hut. After a hearty handshake everyone commenced talking, and we realized that we were once more a united party.

All hands then embarked in the whaler and returned to the ship, where a substantial breakfast had been made ready. We listened to the details of that terrible journey across the plateau, and it was difficult to realize that it was really Mawson who was telling us so quietly of the tragedy of the previous autumn. He looked fit and well, but it was only as one glanced towards the slopes of the great plateau now glistening in the sunlight, that one could picture the awful character of his struggle against the forces of nature on his homeward march of three hundred miles

In the matter of weather, the second winter had proved even worse than the first, and the advent of summer had been hailed with joy. The wireless communication proved a great success during the second year. Temporary stoppages occurred owing to unusual difficulties arising from the constant hurricanes During the summer, for a period of about

THE CROSS ERECTED ON CAPE DENISON. [Photo, Gillies.

DR. X. MERTZ.

LIEUT. B. E. S. NINNIS.

INSCRIPTION ON THE CROSS.

three months, the wireless communication failed on account of the continuous daylight.

It was not long before we were discussing how the remainder of the present season could be most usefully employed. Before finally leaving Commonwealth Bay, Dr. Mawson wished to visit some of the outlying islands, and the Mertz Glacier-Tongue. Some dredging in the adjacent waters would be undertaken, weather permitting, so it was arranged that the cases of stores from the Hut should be got on board as soon as possible, and, after visiting the glacier-tongue, that we should proceed to the westward, sounding and trawling when possible ; at the same time, taking advantage of any opportunity to approach the coast-line.

CHAPTER XVII
AT COMMONWEALTH BAY
For the Third Time—December 14th-25th, 1913

14.12.13. Fine weather in this locality means boat work and plenty of it. The weather was squally during the forenoon, but the anchor is holding well. The wind moderated at 2 p.m. The launch was then put over and ran continuously until 8 p.m., bringing off cases of specimens and miscellaneous stores, which had been stacked near the landing place ready for embarkation.

15.12.13. A south-east gale sprang up this morning. At 6 a.m. the vessel began to drag her anchor, but, after going about half a mile, she brought up and held until 10 a.m., when, in a heavy squall, she broke out the anchor and drifted away into 78 fathoms. With great difficulty we managed to heave in the anchor and chain, and then steamed in close under the ice cliffs and let go again in ten fathoms, where we held on as the wind moderated. Clear, blue sky all day.

16.12.13. We hung on well last night in spite of occasional heavy squalls. The anchor is in ten fathoms, but when we are swinging seaward, we have 20 fathoms under the stern. At 9 p.m. the wind fell light and we swung inshore. On putting the lead over, I found four and a half fathoms over what appeared to be a pinnacle rock. Further inshore the depth increased to seventeen fathoms. We then went off in the launch and spent some time sounding, but could not find less than four and a half fathoms, and that only over a very limited area. Our draught at this time was 17 feet 9 inches. A very nasty berth with the bottom varying from four and a half to forty fathoms in a distance of five hundred feet!

17.12.13. A light north-westerly breeze this morning, and the launch resumed work. At noon light airs from the

STILLWELL ISLAND.
Off Adelie Land.
Photo, Correll.

A PARTY LANDING ON THE MACKELLAR ISLANDS.
Photo, Hurley.

north-west, a clear sky and very hot sun. The ship swung right across the 4½-fathom patch and, apparently, there is no less depth over it, for which I feel thankful.

The weather is extraordinary. A land breeze blows strong from 10 p.m. until about 10 a.m., followed by a light sea breeze or calm for the next twelve hours. Just before our arrival, very strong winds had been recorded at the Main Base for a week, so I hope we may be favoured with *moderate* weather during our short stay in the Bay.

At 8 this evening all the stores had been brought on board, and everyone has been working hard to complete this part of the work.

18.12.13. The land breeze last night did not exceed force "6" and to-day has been very fine with light northerly airs. The launch has been bringing baskets of ice aboard to replenish our water tanks.

This afternoon Dr. Mawson and a small party went to the Mackellar Islands where they will camp for the night. There is a large collection of specimens and other gear on board, much of it being still on deck, so we have plenty to do in getting things stowed. I shall be glad to get away from this anchorage, with all on board safe. It is difficult for any landsman to realize the risks taken in pottering about on an uncharted coast.

19.12.13. This morning we resumed work loading ice. The wind was a light easterly. At noon the wind began to freshen and the launch was sent to bring the party back from Mackellar Islands. They got back all safe and we hoisted the launch. This evening the wind is freshening and there are heavy snow clouds on the northern horizon. It has been more or less overcast all day and the temperature is much lower. Before midnight it was blowing a fresh gale from the south-east, with fine snow. The anchor is holding well.

20.12.13. The snow cleared off at 4 a.m. Since then the south-easterly wind has been blowing with force "8" to "9," with clear weather but partially overcast sky. The ship has been straining heavily at the anchor, which appears to have got a good hold. Dr. Mawson has been packing and settling up his gear

21.12.13. Yesterday evening the wind moderated, and this morning the launch went away dredging at 9 a.m. and later brought off more ice. One tank is now full and the other nearly so. The weather has been sultry all day, and the ice is thawing rapidly about the boat-harbour.

22.12.13. At 10 p.m. the wind died away and we hove up the starboard anchor, which was covered with kelp. The fluke had broken off at the shank. We then steamed down to the black rocks,* about eight miles west of the anchorage. When about one mile distant we sounded in 424 fathoms on a hard bottom. The launch took Dr. Mawson and a party to examine the locality, while the ship steamed about slowly and the trawling gear was erected. At 2 p.m. we sounded in 324 fathoms on mud, and the trawl was then put over. After towing for a short time the net came up full of many forms of marine life and several pieces of rock; the net being slightly torn. We then steamed back to the black rocks and picked up the launch; finally returning to the anchorage and let go the port anchor in twelve fathoms. The weather to-day has been fine throughout.

23.12.13. The launch brought off the twelve remaining dogs and the personal effects of Dr. Mawson's party, so that all is now ready for a start to-morrow. We have been busy all day getting our broken anchor down on deck. It was found that the fluke which had carried away was flawed and only held down by a small piece of sound iron in the centre. This place is an expensive one so far as anchors are concerned, since this is the second one that we have broken, besides losing three owing to the chain parting. Commonwealth Bay is a treacherous locality and one in which you are never really " snug."

This evening there is a strong breeze from the south-east. The launch has been hoisted up in davits and lashed for the night. Weather overcast with snow-squalls.

24.12.13. This morning at 6 a.m. there was a fresh southeast gale, the barometer falling rapidly. The squalls were of great violence. At 8 a.m. the land became invisible owing to the driving spray and drift. At 10 a.m. the wind averaged about 70 miles an hour with squalls of terrific violence. At 11 a.m. it reached the strength of a hurricane. All is ready for slipping cable. Noon: Hurricane, the sea cut off almost flat by the force of the wind. The glass has fallen three-tenths of an inch since 8 a.m. At noon the reading was 28.75".

* The black rocks were subsequently renamed Cape Hunter after our biologist

ICE-COVERED ROCKS NEAR WINTER QUARTERS, ADÉLIE LAND.

[Photo, Hurley

AT COMMONWEALTH BAY

At 1 p m., with a sudden shock, the anchor came home; the stock having broken in a squall of hurricane force. The ship steamed head to wind while the cable was being hove in. Fortunately the weather moderated a bit after this squall and the vessel became manageable. Very high sea—cable and broken anchor hanging from the bow—launch washing out of the davits! We managed to save the anchor and chain, but the launch and forward davit went overboard.

About 4 p.m. the weather cleared a little and we found ourselves off Cape Hunter. Thereupon we stood to the north to clear the western point of Commonwealth Bay It was too thick to see clearly until 7 p.m., when we sighted the Piano Berg. The wind had shifted from south-east to east, so we steamed to the east to bring the sea on to the bow.

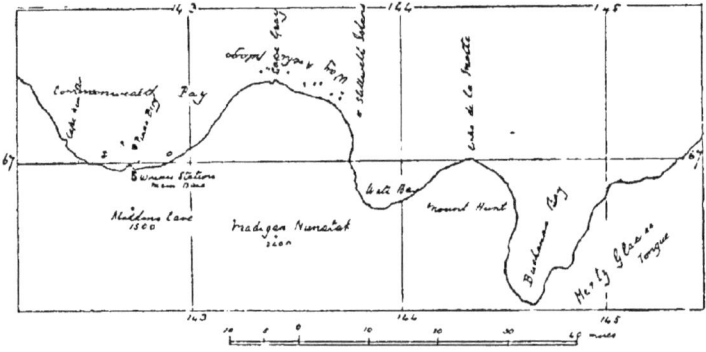

Commonwealth Bay to Buchanan Bay

The weather moderated as we got away from the vicinity of the ice-cliffs. Glass still low, but inclined to steady.

We had a fortunate escape—I do not think a vessel has ever ridden out such a gale as raged from 9 to 12 to-day. I was in hopes that we should be able to hold on right through, as we were well up under the cliff and the wind appeared to be stronger aloft than on deck.

25.12.13. A miserable Christmas Day. Strong wind, confused sea and almost continuous snow. We have been standing to the eastward against wind and sea, making from one and a half to two knots. At noon we were about twenty-eight miles from the Main Base. The barometer was steady at 28.70 when the wind moderated, and the weather cleared a little. Thick snow began to fall about 9 p.m. As our position was uncertain the engines were stopped for the night. Very little floating ice is visible. One berg and a few small pieces of floe were sighted this afternoon.

Our departure from Commonwealth Bay was in keeping with the various troubles we have experienced there Fortunately we have completed our work ; the hut has been fastened up, and we are not obliged to return to this windy spot.

26.12.13. At 8.15 a.m. we stood to the eastward, the weather still remaining thick and overcast. At 3 p.m. the weather cleared a little, so we stood southward and made out the blink of the land. No observations were possible, and the compass error is uncertain The course was altered to E.S.E. along the land, and at 9 p.m. the loom of the Mertz Glacier-Tongue was faintly visible. The snow has been thawing as it fell all day, and everything is damp and chilly. After the last three days a little sunshine will be a pleasant change.

27.12.13. During the forenoon the weather was very thick as we steamed to and fro under the lee of the ice-tongue. At 3 p.m. the weather cleared and we followed the ice-cliff to the north-east until stopped by a line of pack setting round what appeared to be the seaward end of the Tongue, and drifting to west and south-west. The vessel was stopped off a remarkable cave in the ice cliff and a sounding taken in 288 fathoms on mud. We then stood south-west at half-speed.

28.12.13. We took advantage of the fact that the weather was fine this morning by putting over the trawl in 300 fathoms about ¼ mile from the face of the cliff, and towing it for an hour with 500 fathoms of wire. A fine haul was secured, mixed with mud and pieces of rock. The gear worked remarkably well, and the biologists were busy for some time sorting the catch—rather a cold job. Dr. Mawson told me that, among the mud and débris brought up, there was some fossilized wood, which would be a most interesting specimen. We were able to obtain sights for position to-day, and the weather conditions are improving.

29.12.13. About 11 a.m. we were in the same position —off the cave in the cliff—as when we sounded in 288 fathoms on the 27th. The sun came out brightly and a party put off in

THE MERTZ GLACIER TONGUE.

A CAVE IN THE ICE-WALL, MERTZ GLACIER TONGUE.

the whale-boat to inspect the cave. Several photographs were taken of the cave, and also of the ship. We then proceeded to the noith-western point of the tongue until the ship came up to heavy pack. To the north-east a number of big bergs were visible. We were unable to see round the point of the tongue, but everything indicated that we had reached its northerly point on the western side. During the afternoon we worked two plankton nets—one at the surface and the other at a depth of ninety-eight feet.

10.15 p.m. On reaching the entrance to Buchanan Bay, the course was altered to north-west to pass about five miles off the land.

30.12.13. At 1 a m. we were close to a big berg about six miles from the coast. The engines were then stopped until 7 a.m., when we steamed in towards a large rocky island. When about a mile from this island, a party left in the whale-boat to examine it. At noon the sky was cloudless with a gentle breeze from S S.E. Our position was then in latitude 66° 54½′ S., longitude 143° 51′ E. At 12.45 p.m. the boat returned and was hoisted up. The scientists had had a successful trip; a nesting place of the silver-grey petrel having been found, and several eggs obtained.

A course was then set to pass about twelve miles north of the east point of Commonwealth Bay. At 3.30 p.m. we passed the last of the large grounded bergs off the eastern point of the Bay. At 7 p.m. we hauled in the towing nets and proceeded west at full speed. By midnight we were approaching the western point of Commonwealth Bay, when we reduced speed. Many large bergs were grounded off the cape.

31.12.13. The wind freshened during last night, and this morning a fresh gale was coming down the ice-slopes. We steamed about under shelter of the land until 1 p.m., and then, as the breeze did not seem likely to moderate, we bore away N.N.W.

After we had got some distance from the ice-slope, the wind moderated. At 4 p.m. we sounded in 157 fathoms on mud. The trawl was then put over, and by 7 p m. we had it aboard again with a good catch. After sounding, we took a set of serial temperatures and water samples.

The New Year was welcomed with the customary honours by the Expedition members in the Ward Room.

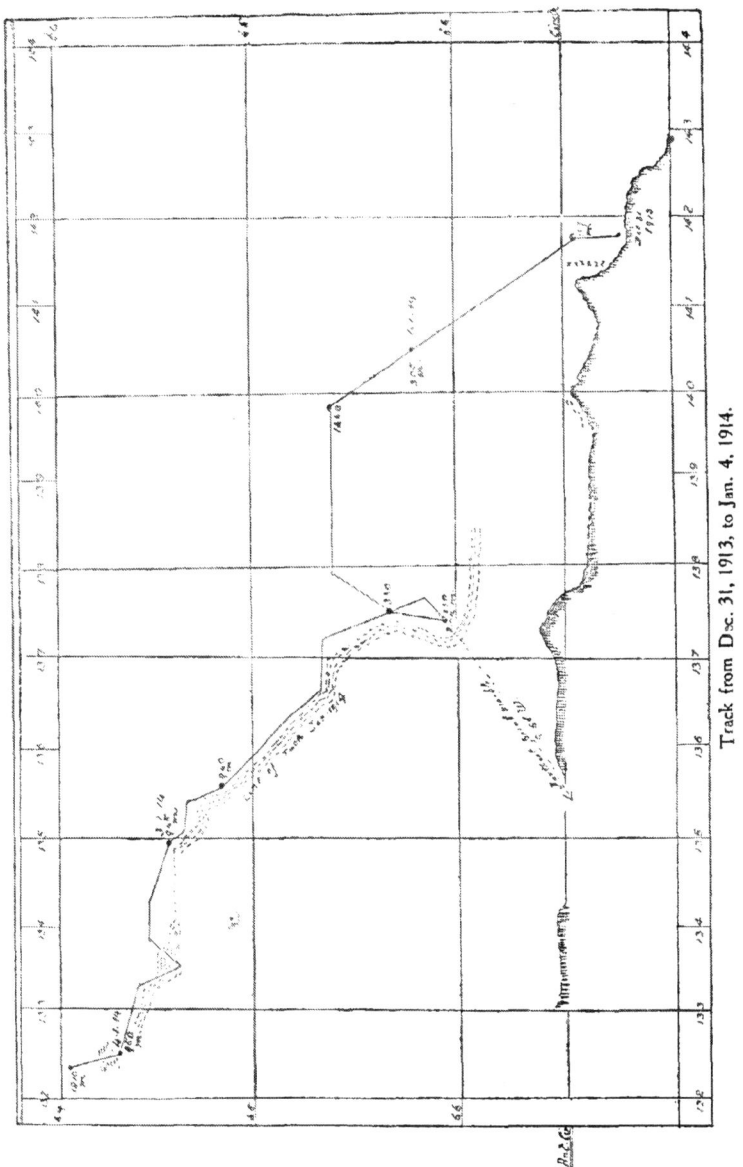

Track from Dec. 31, 1913, to Jan. 4, 1914.

CHAPTER XVIII

WESTWARD HO!

1.1.14. We have made a fair start on our voyage to the west. A fresh gale from the E S.E. has been blowing all the forenoon, our course being N.N.W. true. The glass is steady, and, although many bergs are in sight, the pack-ice is negligible in quantity.

At noon the sun came out and the wind moderated rapidly. We sounded in 205 fathoms, and, at 5.30 p.m., in 1,440 fathoms; this fact indicating an abrupt, continental shelf. The course was then altered to approach the line of pack-ice which we had followed (in a north-westerly direction) for three days, in 1912.

To-night the wind is a light southerly with a confused sea; the sun is setting in a blaze of golden light, the rays being reflected from tabular bergs on the northern horizon.

2.1.14. A very fine day with clear sky and bright sun throughout.

3 a.m. No visible sign of pack, although the ship was some eight miles to the west of the track followed in January, 1912. We then steered to the S.S W., and this course was followed until 6 a.m., when a sounding was taken in 330 fathoms. The ship proceeded on course S. 22 W., and at 7 a.m. the pack was observed from aloft, bearing between south-west and south.

11 a.m. Our further progress was stopped by a heavy line of pack; the land being visible on either bow. Whether this pack extended to the land, or whether open water was present off the land further south, could not be determined.

Noon. We sounded in 250 fathoms on mud and small stones About a mile to the south of our position we could see earth-stained bergs having an appearance very similar to islands. We put the dredge over and paid out 400 fathoms of wire.

After towing for about an hour and a half, the net was hauled in, but the catch was very small, we had been towing into deeper water, as, when the net was aboard, a sounding gave 330 fathoms. We followed a long line of brash-ice to the northward until midnight, when the line trended nearly west and the course was altered accordingly in that direction.

5.1.14. Observations placed our position at noon in latitude 64° 10′ S., longitude 130° 3′ E.

We have been skirting the edge of a huge mass of pack, which is further north than I had expected to find it. The ice is apparently that of last year, and very rotten. We are being driven west, but are still hoping that the pack will take a southerly trend.

At 1 p.m. we sounded in 1,550 fathoms on mud. At 11 p.m we were off a deep bay extending southward in the pack. The ship stood in to the head of this bay, whence was observed an immense field of close, heavy pack. As nothing could be seen beyond it, I came to the conclusion that this field of ice was a part of the main coastal pack. Thereupon the ship was turned round and we pushed northward into open water again.

A bank of cumulus clouds with a very ragged edge was visible all round the southern horizon. This indicated a violent disturbance, probably a gale coming off the ice-slopes, although in our position the weather was nearly calm. These violent gales seldom reach far from the land.

6.1.14. The weather has been unsettled to-day; wind east with light snow, and thick weather at intervals. At 10 a.m. we sounded in 1,700 fathoms on mud, and then shot the trawl; the line of bergs being about two miles to the southward. We paid out 2,526 fathoms of wire, steaming at slow speed before a light westerly wind. At 1.40 p.m. the trawl brought the ship up, and we were obliged to "knock out," to save the gear, as the vessel was riding heavily to it, stern-on. While steadily heaving in, all went on well until about 100 fathoms from the net, when kinks began to show in the wire. All but one of these had pulled out; this one, however, came up like a "turk's head." When we got the net aboard, it was found to be well filled with a varied mixture of mud, pieces of rock, and several biological curiosities, so a long day's work had not been in vain. The derrick-head

A WHALE RISING CLOSE TO THE SHIP. [Photo, Gillies.

WHALE SPOUTING. [Photo, Gillies.

block was slightly injured, but otherwise the gear had stood a very severe test satisfactorily.

7.1.14. We have been steaming all day along heavy pack in which many bergs were marooned. The sky has been heavily overcast, which is favourable for seeing the ice-blink. A huge mass of barrier-ice was observed inside the pack, but as we had passed well to the south of this position in 1912, it must have been a fragment (some 28 miles long) surrounded by the pack. Many whales have been observed spouting during the last few days.

During the next three days our course was westerly, the ship being driven north-west from time to time to clear the pack. Another huge mass of ice was sighted about four miles inside the pack on January 9th. We then sounded in 1,400 fathoms on mud. Some terns were seen flying about on the following day. A great number of water-worn bergs have been visible outside the line of pack since January 9th.

11.1.14. A moderate westerly wind blew to-day; the first from that quarter since we left Adélie Land. The weather is bright and fine. There is an appearance of water sky to the southward, but it is one of those days when the blink is very deceptive. We shall be in the longitude of Knox Land to-morrow.

12.1.14. The wind has been freshening from the east, with a falling barometer and rising sea since 9 a m. At that hour we sounded in 1,530 fathoms on mud. At noon our position was in latitude 64° 37′ S., longitude 108° 50′ E., and we followed the edge of the pack closely, until the evening, looking out for any lead south in the direction of Knox Land. Thick weather came on, so the course was altered to north-west true, and the engines were reduced to half-speed. About midnight it was blowing a fresh easterly gale with snow-squalls, the vessel shipping water fore and aft.

13.1.14. At 4 a.m. the ship was running to the westward before the gale. Three large bergs were sighted The same weather conditions lasted throughout the day; the sun showing through the clouds at intervals. At 6 p.m the pack was sighted, and both wind and sea were moderating. It was observed that the pack was becoming very dense—a consolidated mass of small bergs and heavy floes fringed by smaller bergs.

At 10 p.m the course was N. 15. W., glass rising, and wind moderating rapidly. The movements of the barometer indicate that the ship has been in the southern semi-circle of a cyclonic disturbance which appears to have passed some distance to the north of our track.

14.1.14. The weather has been fine all day.

At 9 a.m. we sounded, and found that the water had shoaled to 710 fathoms. After steaming seven miles to the north, we sounded again in 870 fathoms on mud. This

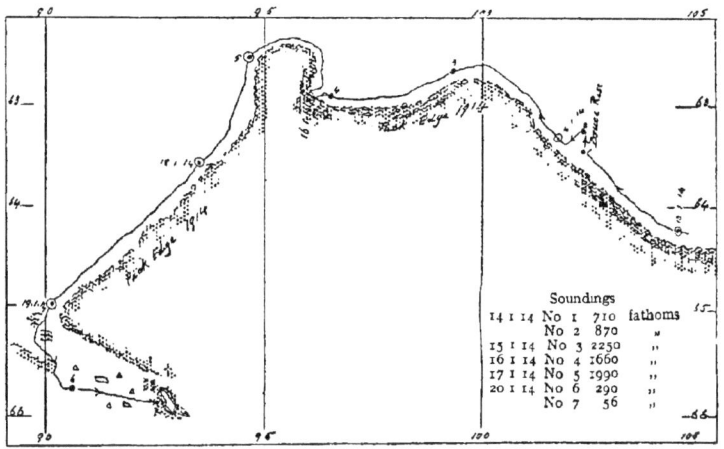

Track of "Aurora" from 13 1 14 to 21 1 14, when the ship was off Drygalski Island in 56 fathoms on hard bottom

shoal was named by Dr. Mawson in honour of Dr. W. S. Bruce—the veteran Polar explorer—the Bruce Rise. The trawl was put over and towed for half an hour with 1,100 fathoms of wire. The catch included stones, pieces of volcanic rock, two deep-sea fishes, and a number of red prawns. A temperature section was then taken.

We are being driven a long way to the north by this dense line of pack, the edge is somewhat broken up by the recent gale.

15.1.14. We were forced to steer to the north last night by the pack, which was fringed by a number of bergs of various types. The "Aurora" passed close to one huge, pinnacled berg, which measured by sextant angle, was 200 ft. high. It was a grand sight!

We were then able to steer west about 4 a.m., and, during the forenoon, we sounded in 2,250 fathoms on mud. The weights failed to disengage and were recovered from this depth. The pack is trending a little south of west. During the afternoon the weather was thick, with heavy snow-squalls.

16.1.14. This morning at 6 a.m. we stopped about a cable's length from the line of pack and sounded in 1,660 fathoms on mud. The weights again came up with the driver. After breakfast we entered the pack-ice and stood south Two crab-eater seals were shot and hauled on board, and we worked slowly through the floes. At noon we were about two miles inside the line of pack; the weather being fairly clear. The ice was then becoming too close for further progress, so we tied up to a large floe and proceeded to water ship. This occupied about seven hours, and, during this time, we observed that the floes were drifting steadily to north-west true. When we were making our way out, it was found that the pack had closed up a good deal, rendering it difficult to reach the open water again.

11 p.m. We are now steaming along the edge of the pack, which is trending to the north. It is nearly calm and the moon has just risen, shedding a golden light over the white bergs and dark water; a very fine picture.

17.1.14. At noon we had reached the northern extremity of the pack, when the course was altered to south-east true, until 4 p.m , when we sounded in 1,990 fathoms on mud. The bottom-temperature (recorded by the thermometer in float) was $2°.17$ C.

The weather has been remarkably fine, the sea smooth, and the sky overcast. After sounding we steered south true, passing earth-stained bergs. At 10.40 p.m. we met the pack again and were headed off to W.S.W. true.

18.1.14. This evening we are running through loose pack on a south-westerly course. The further west we move from Termination Barrier, the better the ice conditions become. A great many Finner whales have been about for the last few days; some of which have been of exceptional size. There appears to be some heavy pack ahead, but, in a very bad light, what appears to be dense pack frequently turns out to be some pieces of scattered ice miraged up.

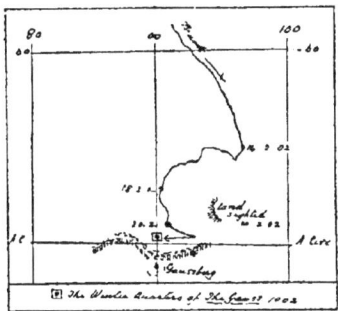

19.1.14. We came through a wide expanse of scattered pack last night; to the eastward it was closer, and to the westward clearer. All the ice appeared much decayed, but heavier than that which we came through last month off Commonwealth Bay.

At 9 p.m. open water was observed ahead extending round the horizon from W.S.W. to east. We were then about thirty miles N. 30 W. of the position where the "Gauss" was frozen in.

A course was then set for Drygalski Island, which is about eighty miles to the eastward. The water was studded with large bergs, but the pack had been left behind, for the time being.

CHAPTER XIX

QUEEN MARY LAND

DRYGALSKI ISLAND, TO THE SHACKLETON ICE-SHELF

20.1.14. Noon. We have been steaming east since 10 p.m. yesterday through fairly open water with bergs about. There is an appearance of pack to the northward.

Drygalski Island, from a sketch made by A. J. Hodgeman, 21.1.14.

The island rises to a height of 1,200 ft. in the centre.

The overhanging top. Average height of vertical cliffs, 165 feet.

At 1 p.m. Drygalski Island was sighted, also the high land to the southward. We continued to approach the western coast of the island until 9 p.m., when the ship was about three miles from the shore. Vertical ice cliffs rose from the water's edge; not a vestige of land being visible, although the water had shoaled from 265 fathoms on mud to 56 fathoms on a hard bottom. The engines were stopped for the night, as we had decided to put the trawl over in the morning at 6 a.m. The island is ice-capped, appears to be about nine miles long and rises to a height of 1,200 feet in the centre.

21.1.14. The vessel had drifted slowly during the night, and, after we had got into position again (with a depth of 54 fathoms), the net was put over with 100 fathoms of wire. The bottom proved to be good, as, after towing for half an hour towards the island, we brought up the trawl nearly full. It contained a varied collection of marine life, without any rock or mud.

We steered towards the northern point of the island, but heavy pack prevented us from rounding that end. We then turned and steered in a southerly direction along the coast, taking frequent soundings, which averaged about 56 fathoms until the ship had reached the southern extremity of the island, where the water deepened to 100 fathoms. The position of the southern point of the island is in latitude 65° 49′ S., longitude 92° 30′ E. (approximate).

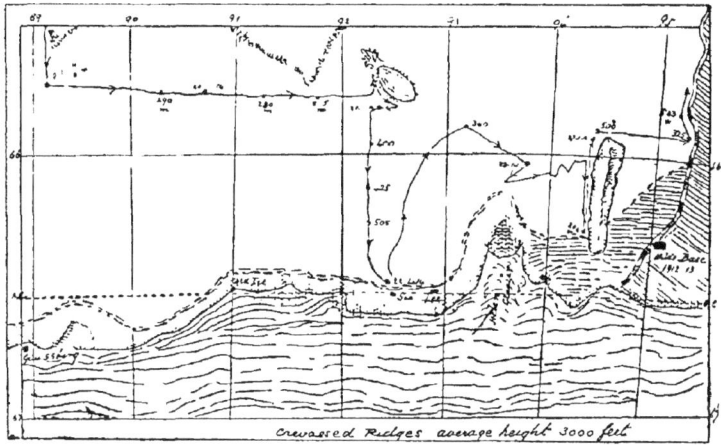

Track of the "Aurora" off Queen Mary Land, Jan 19–Jan 27, 1914

Vertical cliffs rise from sea level to an average height of one hundred and sixty-five feet. The dome-shaped ice-cap rising above them to 1,200 feet in the centre and gradually falling away to north and south, is a remarkable sight. The position of this island in a line of shallow water is very striking, and may suggest an explanation of the immense ice sheet which covers it.

At noon we started south at full speed, and, two hours later, the water had deepened to 400 fathoms. We followed this course until 8 p.m., when we were some eight miles distant from the inland ice, which formed a bay fringed with fast sea-ice of considerable thickness. Rookery Island lay to the

[Photo, Gray.
PLACING THE BLOCKS OF ICE IN TANK THROUGH WHICH A STEAM COIL P

[Photo, Hurley.
WATERING SHIP FROM THE FLOES.

eastward; a large mass of black rock which appeared to be the end of a reef extending from the shore. This island, with no snow on it, contrasted strongly with the white bergs and the ice-covered land. We steered toward it until stopped by fast ice extending round the horizon to south and east. The island was then at an estimated distance of from six to eight miles; it was not until the ship was within four miles of the fast ice, that we realized that Rookery Island was not surrounded by water. The black rock was elevated by refraction, and the fast ice, although much closer to us, was not seen until later. It was a glorious night when the ice anchor was put out, everything was illuminated by the soft reddish light from the sun setting behind some cirrus clouds. To the south, the snowy slopes showed indistinctly, and the bergs all round us in the shadows looked like huge monsters in a field of white. There was a noticeable movement in the water and large pieces of ice were drifting away to the northwest.

22.1.14. The weather continued fine, and the ship was moored to the floe fore and aft while we got some ice aboard. Many killer whales were observed round the ship Seven Ross seals were secured, and some Emperor penguins. Weather fine, but overcast.

At 2 p.m. we got under weigh on a N.25 E. course. At 11 p.m. we sounded in 360 fathoms; barometer falling, strong easterly winds and rising sea; course altered to south-east.

23.1.14. Strong gale from the east, ship steering badly; bergs and drift ice about; making about south-east.

24.1.14. We are having a bad time; high sea; thick snow, and innumerable bergs. We have been trying to reach the big barrier berg off Wild's base.

10 p.m. We have just made the barrier berg near its northern end. Once under its lee, we were in smooth water and could listen to the howling of the wind with less apprehension.

25.1.14. Still blowing very hard. The wind has remained east throughout. It has been snowing very heavily all day; the weather was so thick that there was great difficulty in seeing the face of the berg.

26.1.14. We have been steaming under the berg all day wearily waiting for better weather. At 8 p.m. the weather had moderated considerably We were at this time off the southern end of the berg where it trends to the east and joins the floe-ice. Apparently, the recent gale had had little effect on this bay-ice, which was further north than when we visited the locality in 1912. We sounded in 204 fathoms on mud, and then turned back to the northern end at slow speed. We passed a low place on the western face of this berg (or glacier-tongue), where the cliff was not more than thirty feet high The snow slope was visible running up from this curious break in the line of cliff. The ice was of an unusual type, more like a kind of ice concrete, with a great deal of blue ice.

27.1.14. A fine day with light east wind and clear weather; a delightful change after four days' blow! The dredge was put over in 140 fathoms on a rocky bottom. After towing it a mile to the north, we heaved up the trawl—or rather, what remained of it, namely, the top cross bar to which the bridle is made fast; the net was *lost*. We then put over one of the small dredges with 140 fathoms of wire and obtained a good haul.

At noon we were about nine miles from the northern point of the berg (in latitude 65° 54′ S.), under which we have been sheltering. I am inclined to think that this mass of ice is a remnant of a bigger glacier supported by an island in the centre, rather than a big berg After sounding, we steered due east until 8.40 p.m. when we reached the eastern point of a wide bay in the Shackleton Ice-Shelf. We sounded about a mile from the cliff in 328 fathoms on mud. This sounding was taken about five miles S.S.E. of the position where we had found 543 fathoms on hard bottom in 1912.

To-night is overcast but calm, with a little snow at intervals.

10 p.m. We are now following the ice cliffs of the Shackleton Shelf to the north at half speed.

28.1.14. Fine weather with light northerly winds; low mists coming down at intervals. This morning we sounded in 225 fathoms, but no sample was recovered. Two hours later we sounded in 240 fathoms on mud. The trawl was put over and towed for twenty minutes, securing a good catch.

At 4.30 p.m. we were off the end of the Shackleton Shelf, further east than we had been before. A huge mass of ice was sighted to the northward, and what appeared to be

[Photo, Correll.
PUSHING THROUGH PACK OFF TERMINATION TONGUE.

[Photo, Mclean.
A "WATER SKY" IS INDICATED BY DARK LINE ON THE HORIZON.

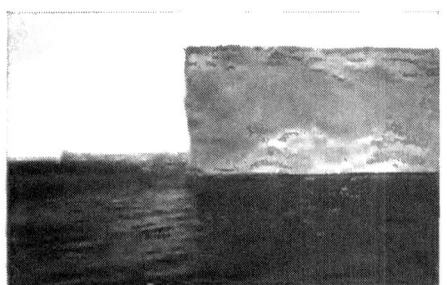

[Photo, Davis.
EDGE OF TABULAR ICEBERG OFF SHACKLETON SHELF.

[Photo, Gillies.
OFF TERMINATION TONGUE.

[Photo, Gray.
LOOSE PACK OFF TERMINATION TONGUE.

heavy pack to the west. The light being bad, these observations were indistinct. However, there was clear water to the E.N.E. and we followed that course until 9.30 p.m., when we reached some of the heaviest fast ice I have ever seen. It appeared to fill the head of the bay between the Shackleton Shelf to the South, and the long mass of ice to the North, of which, owing to thick weather we had not been able to get a clear view; low misty clouds shut out everything.

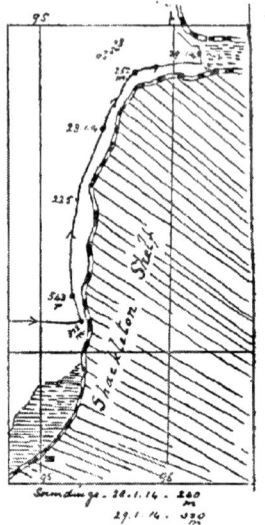

Track of the "Aurora."

We lay-to for the night waiting for the mist to clear. The barometer reading at midnight was 29.47″.

29.1.14. The weather continued misty until 5 p.m. this evening, when we steamed up to the fast ice a few miles to the north of where we were last night. We found the surface covered with three feet of soft snow, in which we buried an ice anchor and a large plank which held satisfactorily, the wind being very light.

A few islands were seen at some distance inside the fast ice, and Dr. Mawson went over the floe to get a nearer view.

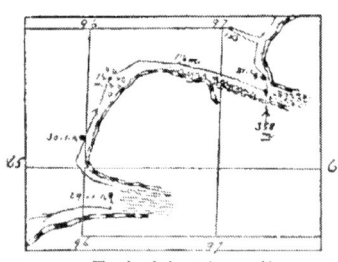

Track of the "Aurora."

On his return he told us that what looked like islands were only bergs frozen into the ice. As well as I could see from aloft, the line of ice cliffs that we have been following fades away in the south-east and there is an appearance of open water in that direction. Behind this fast ice, and subtending nearly six compass points, a line of barrier appears to commence, trending about W.N.W. and gradually curving to the north. A balloon would have been a great acquisition, but, as one was not available, we could only plot the apparent position as seen from the crow's nest.

In the distance, refraction and the shadows cast by the western sun distort everything; for example, the appearance of open water may be a shadow of an ice cliff on a distant floe.

30.1.14. Our ice anchor held well all night, and we proceeded to follow the ice cliff to W.N.W. at 9.30 this morning.

When within about 300 yards of the ice face, which was from sixty to seventy feet high, an immense mass of ice broke away from the face of the cliff, and sank into the sea; it rose and sank alternately during the next few minutes, breaking up rapidly, but without noise. At the end of five minutes, two blue bergs were the most prominent features in the debris resulting from the collapse. The sea was disturbed for perhaps a quarter of a mile; less than I should have expected from the sudden fall of about four hundred feet of the cliff face.

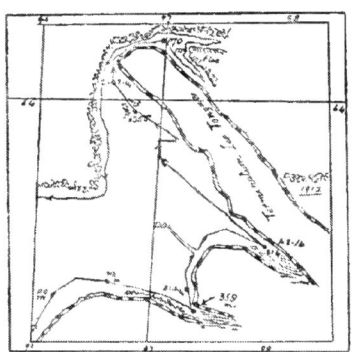

Track off Termination Ice-Tongue.

At noon we sounded in 370 fathoms on mud (latitude 64° 53′ S., longitude 95° 59′ E.). The barrier face is lower than that we followed on January 28th, and is now trending north-east. At 2.30 p.m. we reached a position where an immense collection of bergs hid the ice-cliff from view. Sounded in 110 fathoms. The bergs are of various shapes and appear higher than the ice-cliff. They all show a curious tide mark, which for some three feet above the water-line is smooth ice. Whether this is due to tilting or to tide is uncertain.

At 5.30 p.m. the sky cleared suddenly and the sun came out brilliantly. Sounded in 114 fathoms on mud. The ice-cliff could be seen occasionally through the bergs with a fringe of fast ice attached to it. As the bergs became less numerous, we were able to steer S. 75 E., and this course was followed until 10 p.m., when fast ice was observed right ahead.

At 10.30 p.m. we were stopped by the fast ice, which was not more than nine inches above water, and without snow. A number of bergs were frozen into it and the ice showed little

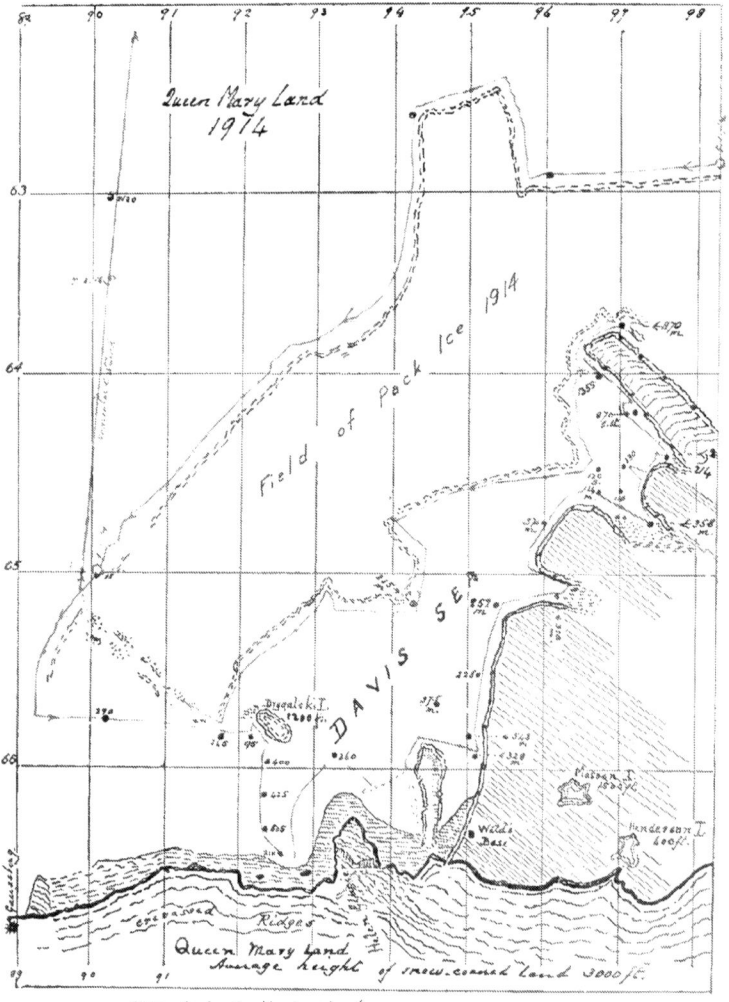

Queen Mary Land, 1914.

sign of breaking up. We put the ice anchor out and sounded in 358 fathoms on mud Our position by stellar observations was, at this point, in latitude 64° 44′ S., longitude 97° 29′ E.

31.1.14. At 9 a.m. a party of four landed on the floe to get some seals and to take photographs. The ship then left the floe to dredge in the bay. Weather calm. The trawl was shot in 358 fathoms, but the catch was very poor; the bottom was evidently very soft mud. At 11 a.m. we picked up the party from the floe, and followed the barrier to the northward.

At 3 p.m. we sounded in 111 fathoms, and, immediately afterwards, put the trawl over, but the net came up empty; apparently it had not reached bottom. We sounded again on hard bottom, and let go the trawl with 230 fathoms of wire. This effort was successful, as the net came up *full*; an excellent catch. At 8 p m. the deck was cleared up, and we took a series of water temperatures and samples. The ship was then allowed to drift slowly to the north-west.

1.2.14. After drifting to north-west during the night, we sounded at 6 a.m. in 130 fathoms, and then made the position off the barrier where we had been at 8 p.m. yesterday.

At 10 a m. we were crossing the bay where the shoal water appears to end. We then stood south-east between bergs, into a deep bay, similar to the one we were in yesterday. At noon we were up to fast ice, which was wind-swept, and polished. There was a break in the ice-cliff; the southern face appearing to end abruptly at less than three miles from our position, while the northern face disappeared to the south-east at a long distance off. It *appeared* to lose itself almost directly behind the nearer face, on a bearing of S. 60 E true, at an estimated distance of fifteen miles. Nothing was visible between the bearings of the two faces (using a powerful telescope), except smooth, floe ice, with an occasional berg frozen in. The sounding at noon, close to the edge of the fast ice, was 214 fathoms.

During the afternoon we steered approximately north-west and obtained good sights for longitude at 5.17 p m (97° 6′ E). There was a strong ice blink to westward denoting the presence of pack, which was sighted shortly afterwards. At 5 p.m. we headed up to the face of the cliff. An hour later the wind went into south, and fog came down on us.

L

At 7.30 p.m. pack was visible ahead. The engines were stopped, and a sounding was taken in 1,355 fathoms about a mile from the western face of Termination Ice-Tongue. A thick mist continued until 11.30, when it cleared a little, and we could just make out the towering cliffs—much too close to be pleasant.

At 11.45 a golden glow lit up the sky to the south-west. The mist cleared off as if by magic, and the stars shone brightly; the change was as welcome as it was wonderful, Termination Ice-Tongue being an uncomfortable lee shore in thick weather. At midnight, good stellar observations gave the position as latitude 64° 2′ S., longitude 97° 2½′ E.

2.2.14. We stood some distance to the south last night in order to keep well to windward of the barrier. At 8 a m. to-day we started full speed towards the north-western point. Loose pack was met some distance from the point, and we could see very dense pack all round to the seaward.

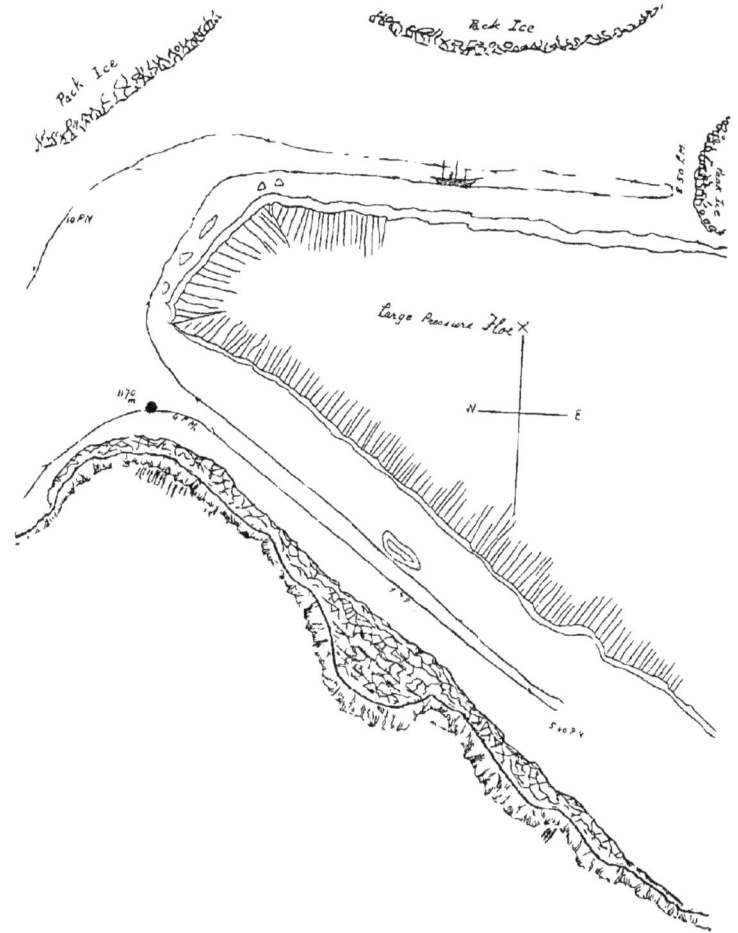

Course of *Aurora* between 4 p m and 10 30 p m , 2nd Feb , 1914 from rough sketch by J H Blair, Chief Officer   The cliff of North Eastern face of Termination Tongue varied in height from 40 to 60 feet   Peculiar, fast ice fringed the foot of the cliffs   The large Pressure Floe was about 6 feet above the water   Approximate scale 2 miles to 1 inch.

CHAPTER XX

THE HOMEWARD VOYAGE

2.2.14. At noon we were 5 miles from the north-western extremity of Termination Ice-Tongue; the pack setting on to the foot of the ice-cliff, so we had to steam at full speed to avoid being caught between pack and barrier. We met some heavy floes just off the point, but with a bump or two, we managed to get through a passage about half a cable wide between pack and barrier. Here the Tongue trended sharply to E.N E., and we were able to steam through open water close to the ninety-foot cliff. The pack to the northward was lost to sight until 4 p.m., when it became visible again. At that hour, the ice-cliff was trending south-east by south true. We sounded in 1,170 fathoms on mud (latitude 63° 47′ S., longitude 96° 58′ E.). All the way along the northern face of the tongue there was a peculiar fringe of fast ice, which sometimes stood out from the foot of the cliffs as much as 400 yards. It was not flat, but most irregular on the surface; like a heap of compressed sea-ice.

After taking a sounding, we steered down a channel about two miles wide, to S. 54° E. true, until 5 40 p.m., when open water was still visible ahead, but heavy pack was present to the north and east. As there was some risk of being caught between the Tongue and the floe, if we continued on this course, Dr Mawson decided that we should turn round, and at 7 p.m. we were close to the place of sounding. We then steamed east, down a line of fairly open water, with an unbroken floe of heavy pressure-ice to the south. After going about ten miles on this course, we turned round and regained the northern end of the Tongue about 11.30 p m Weather thick, with snow, so the engines were stopped to wait for daylight

3.2.14. At 4 a.m. we directed our course towards the north-westerly point of the barrier, but owing to thick weather we had several " halts " for longer or shorter periods. At 11 a.m. we had reached our objective, and in less than half an hour we had made open water to the west of the Ice Tongue, and were steaming S.S.W.

As we were rounding the north-westerly point, I noticed (1) that the pack through which we had pushed yesterday had been

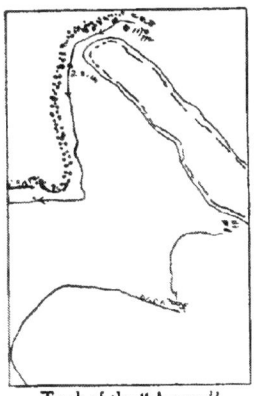

Track of the "Aurora."

replaced by two of the big pressure floes which had set in from the north, and (2) that a rapid current was settling round the point, and streaming away to N.N.E., carrying pieces of ice with it. At 2 p.m. the weather cleared up, allowing us to follow the edge of the pack to the southward until 5 p.m. when it was found to be trending to west true.

9.15 p.m. The sun has just set, leaving the western horizon a light golden colour, and a few small cumulus clouds to the southward a deep crimson.

4.2.14. A fine day with a moderate south-westerly wind. We have been coasting along heavy pack, making a little westing at times. This evening, our course is about W. by N. true. The Chief Engineer told me to-night that the condenser was leaking.

5. 2. 14. At 8 a.m. we sighted Drygalski Island. As the weather was fine (at noon), with a very light wind, the vessel was stopped to the west of the island for engine-room repairs. When things go wrong below, it is well to get them fixed up during fine weather. At 4.50 p.m. we started again, and two hours later entered loose pack on a N. 64° W. course, continuing to push through at various speeds until 11 p.m., when the engines were stopped till daybreak.

6.2.14. We continued our way through fairly open pack and got on well for the next six hours. Then the pack became much closer; we steamed on as fast as possible, as there was a strong appearance of open water to the north-west.

[Photo, Davis.
OFF TERMINATION BARRIER TONGUE. ICE BLINK OVER BARRIER.

[Photo, Gillies.
CLOSE VIEW OF PORTION OF TERMINATION BARRIER TONGUE.

Soon afterwards a faint swell was perceptible. At noon the weather cleared and the sky was almost cloudless, but the water-sky disappeared with the clouds. At 2 p.m. open water or a large lead was observed from aloft by Gray (the officer on watch) to north-west true ; we stood in that direction, and the swell increased rapidly as we advanced.

When we did reach open water at 5.30 p.m., it was almost in the same longitude as that in which we had passed the line of pack coming south (89° 56′ E.). From this, it would appear that the mass of pack had scarcely shifted its position, despite the four days' heavy gale that we encountered after we had got to the south of it. Probably the westerly swell, which is so marked here, balances any tendency of the pack to drift to the west.

At 6 p.m. we were clear of all pack and steaming full speed on a northerly course.

7.2.14. I am glad to be out of the ice and in the open water once more ; the strain has been very heavy for the last three weeks. The members of the Expedition are all anxious to get home, and I think Dr. Mawson has come to a wise decision in not risking being caught in the ice by prolonging the present voyage. I feel very thankful that everyone is safe and well on board, bound for home at last, after spending two years in Greater South Australia.

We sounded in 2,120 fathoms on mud. The weights failed to disengage and came up with the driver. This sounding took just one minute over the hour.

8.2.14. Several bergs about to-day. All sail was set for the first time since we entered the ice.

The last considerable amount of ice was sighted on February 12th in latitude 56° 17′ S., longitude 95° 27′ E. ; one solitary berg being seen later in latitude 55° 9′ S.

The voyage to Adelaide was completed on February 26th ; we berthed at Port Adelaide, having eighty tons of coal on board. A few soundings taken during this part of the voyage are shown on map at the end of this volume.

Our welcome to Adelaide was of characteristic Australian warmth. Numerous receptions were given to celebrate the safe return of the Expedition.

The " official welcome " was held at the University of Adelaide, and was attended by His Excellency the Governor-

General, Lord Denman; Sir Samuel Way, the Chancellor of the University, the Hon. L. E. Groom, Minister of Trade and Customs, who represented the Commonwealth Government, the Hon. Mr. Peake, Premier of South Australia, Professor Orme Masson, representing the Antarctic Committee, and many other citizens of South Australia.

It was an unforgettable scene and one which I am sure will remain a very pleasant recollection to those who were privileged to have had a share in the work of Australia's first Antarctic Expedition.

CHAPTER XXI

THE COASTAL LINE OF PACK-ICE

(1) IN THE VICINITY OF TERMINATION TONGUE. (2) FROM LONGITUDE 135° E. TO 105° E. (3) FROM LONGITUDE 138° E. TO 132 E. SOUNDINGS ALONG THE EDGE OF THE PACK.

AFTER three voyages to the westward from Commonwealth Bay, I have been struck by the remarkable variation in the northern limit of the ice in successive years close to Termination Ice-Tongue. This is shown on sketch overleaf. The northern end of this glacial tongue is about 160 miles from the actual coast-line of Queen Mary Land, and, in 1914, the northern limit of the pack ice was some 200 miles north of the coast-line of the continent. This shows the necessity of determining the coast-line—which is permanent—as distinct from the fringe of pack-ice, which, in this locality, varies from year to year.

With reference to the limit of the pack-ice to the east of the Mertz Glacier-Tongue, the exploring party, led by Madigan, who travelled over the sea-ice late in the summer of 1912, observed the coastal line of pack-ice extending about thirty miles to the north from the Ninnis Glacier-Tongue. The floe-ice is retained in position by glacier tongues, and is thus prevented from breaking up and drifting north, as it does in the Ross Sea area. In several localities, shallow water with numerous grounded bergs check any tendency there is for the ice to break up.

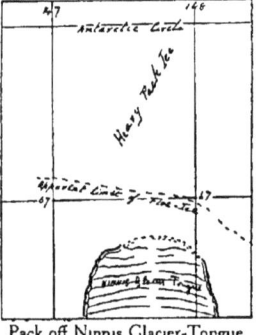

Pack off Ninnis Glacier-Tongue

Between the meridians 135° and 105° E, the northern

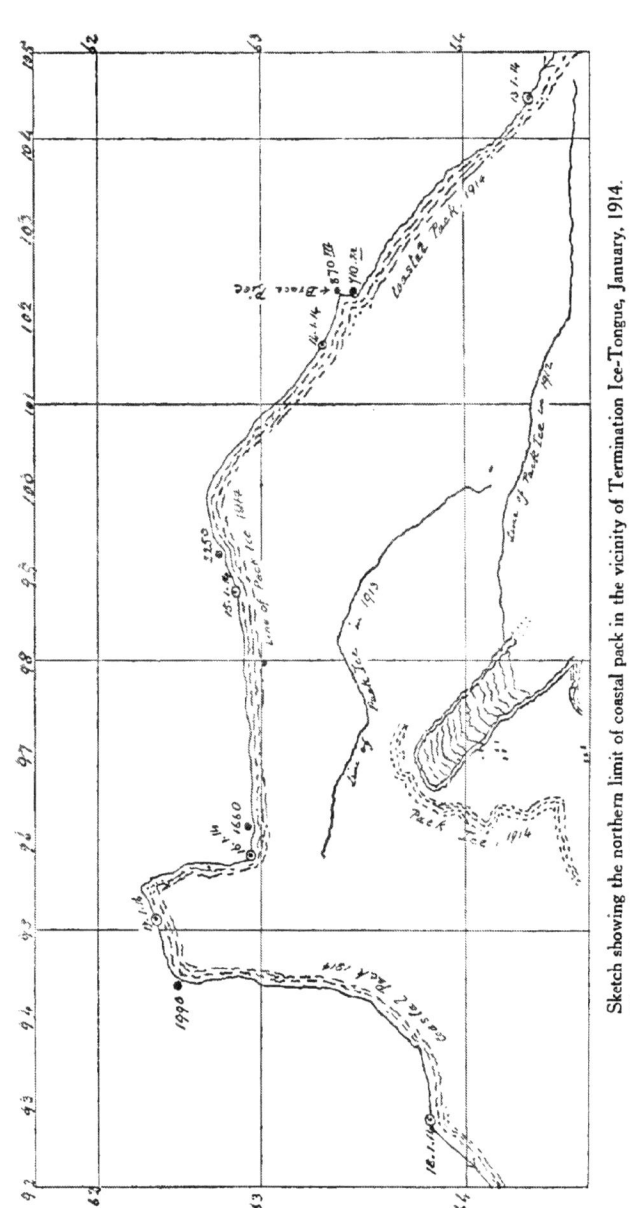

Sketch showing the northern limit of coastal pack in the vicinity of Termination Ice-Tongue, January, 1914. The northern limit of line of pack in 1912, and also in 1913, has been marked to facilitate comparison.

limit of the pack showed little change in successive years but in 1914 it was further north—between longitude 135° and 125°—than it had been in 1912.

The accompanying diagram shows approximately the northern limit of the coastal pack-ice in successive years between the meridians 135° and 105° E.

When following the edge of the pack from longitude 135° E. to the westward, we seldom experienced violent winds. There was a great deal of snow and gloomy weather, but comparatively little wind. This pointed to the fact that the belt of pack to the north of the fast sea-ice had a moderating effect on the violent winds which sweep down from the ice-covered land.

Along this coast, wherever an ice-tongue extends some distance from the land in a northerly direction, the move-

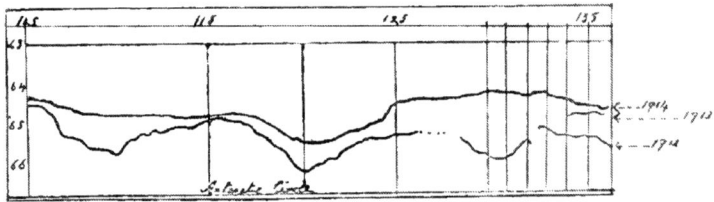

Northern limit of coastal pack-ice.

ment of the pack-ice towards the west is arrested. Open water or loose pack is usually found to leeward or west of a barrier's face, as at Mertz Glacier-Tongue and at Termination Ice-Tongue.

A remarkable field of pack-ice, first reported by d'Urville, extends in a north-westerly direction from the coast of Adelie Land for a distance of about 150 miles. In 1912 the "Aurora" found open water to the west of this field of pack, and we sailed nearly due south in longitude 132° 30′ E. to within ten miles of the coast-line—now called Wilkes Land. The eastern edge of this belt of pack was found *very nearly* in the same position as reported by the French navigator seventy-two years previously.

The diagram (facing the next page) shows the eastern edge of the pack and the track of the "Astrolabe" from January 25th to January 27th. The ship lost sight of the edge during a heavy gale. Meeting it again on January 27th, the ship was driven to the north and did not again encounter it until January 29th. All day on January 30th d'Urville followed an ice wall to which he gave the name of Côte Clarie. This barrier may have been the face of a shelf-ice formation—formed in the same way as the Shackleton Shelf—and the belt of pack-ice which he had followed would have gradually accumulated on the eastern and north-eastern faces of this ice-shelf. This is merely a suggested explanation of the existence of the long belt of pack-ice in 1840.

The lower portion of diagram shows (on the same scale) :—

(1) The edge of the pack followed by the "Aurora" in 1912, marked by a single line—the track of ship in *red*—and

(2) The pack edge in 1914, marked by a *double* line and the ship's track in *black*.

In 1912 we found that Côte Clarie had disappeared, and its former position was occupied by a number of very large bergs in fairly open water. We were able to sail south (as shown), and discovered a long line of ice-covered land—to which the name of Wilkes Land was given by Dr. Mawson.

At noon on the 22nd January, 1912, a sounding of 239 fathoms on hard bottom had been taken. In the same latitude, on the following day, the depth was 160 fathoms on sand and small stones. This sounding of 239 fathoms was the only one on hard bottom obtained in the vicinity of this pack-belt in 1912. In 1914 the eastern edge of the pack-belt had shifted to the westward, about thirty miles south of the 65th parallel. The coast of Adélie Land could be traced to near the 135th meridian, but the existence of open water off the coast to the south of the pack could not be determined. To the north of the 65th parallel there was little change in the direction of the edge, but on reaching longitude 132° 30′ we found the border of the pack extending to the north, and a good deal of loose ice all around. No open water was visible to the southward. A sounding gave 950 fathoms on mud. The pack was followed to latitude 64° S., when we sounded in 1,810 fathoms on mud. Shortly afterwards we were able to follow the edge of very heavy pack in a westerly direction.

It is not easy to account satisfactorily for this belt of pack—extending in a north-westerly direction from close to the Antarctic Circle to latitude 64° N.—being held in practically the same position for the last seventy-two years. The ice-

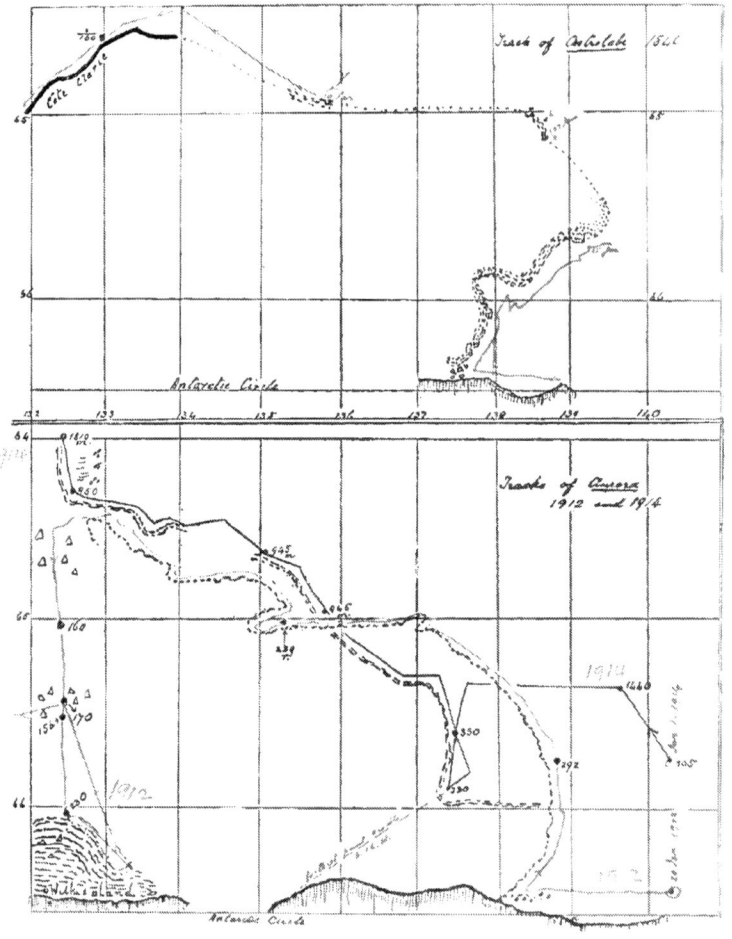

Track of the "Astrolabe," 1840 and the "Aurora," 1912-1914.

THE COASTAL LINE OF PACK ICE

shelf—if such existed in 1840—has vanished. What holds the pack *now*? I can only record personal observations and leave the question to be discussed by scientists.

The sketch shows approximately the northern edge of the pack-ice in 1912 between Termination Tongue and Adelie Land.

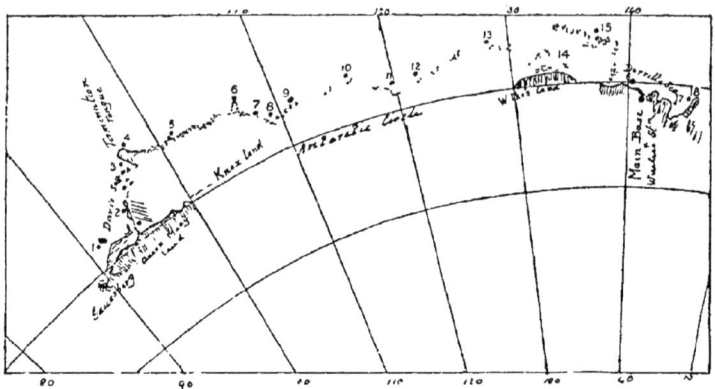

The positions indicated by small dots and numbered 1 to 18 refer to soundings taken by the "Aurora."

Date.
1. Off the coast of Drygalski Island, 56 fathoms, on a hard bottom. . . . . . 21.1.14.
2. Off Shackleton Shelf, 240 fathoms, on mud. 28.1.14.
3. Off Termination Tongue, 1,355 fathoms on mud 2.2.14.
4. Off Termination Tongue, 1,170 fathoms on mud 2.2.14.

The following were taken in 1912 :—

5. 1,080 fathoms lost.	12. 630 fathoms on rock.
6. 1,500 fathoms on mud.	13. 230 fathoms on mud.
7. 450 fathoms on rock.	14. 170 fathoms on mud
8. 300 fathoms on mud.	15. 239 fathoms on rock.
9. 1,150 fathoms lost.	16. 308 fathoms on mud.
10. 927 fathoms on mud and small stones.	17. 32 fathoms —
	18. 398 fathoms on mud.
11. 340 fathoms on mud.	

Some notes on the Antarctic soundings of the "Aurora" were published in the *Geographical Journal*, October, 1913, contributed by the late Sir John Murray.

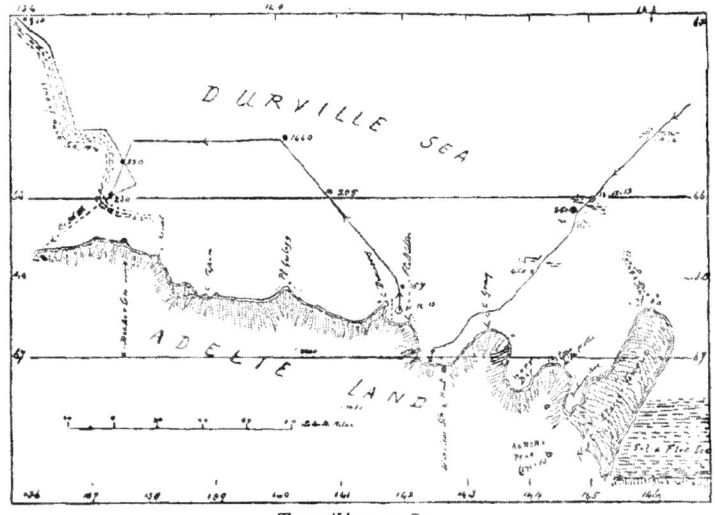

The d'Urville Sea.

CHAPTER XXII

THE D'URVILLE SEA

Open water. The surface current near Termination Ice-Tongue. The Ice Conditions off Queen Mary Land. Soundings in the Davis Sea

As we approached the d'Urville Sea in 1912, heavy pack and an ice wall were met in the position shown on the diagram. We sailed to the southward, and then west until reaching the anchorage off the Main Base.

Twelve months later we were able to approach the Main Base by the track shown. The pack was very loose and the solid ice had vanished.

On leaving the Main Base in February, 1913, to relieve Wild's party, we found pack to the north of Commonwealth Bay, and, a little further north, we steamed along the face of an immense berg for over forty miles. This had evidently formed part of the vanished ice wall.

In 1914 we had very little trouble in approaching the anchorage. With the exception of a few bergs and scattered fragments of floe-ice, I may say that d'Urville Sea was unimpeded by ice in the summer of 1914.

The sea surface near the coast is kept free from ice by the violent southerly winds. Gale succeeds gale so rapidly that ice of sufficient thickness to resist the violence of the wind has no time to form. It has often been noticed that when a strong blizzard is blowing off-shore, there is little

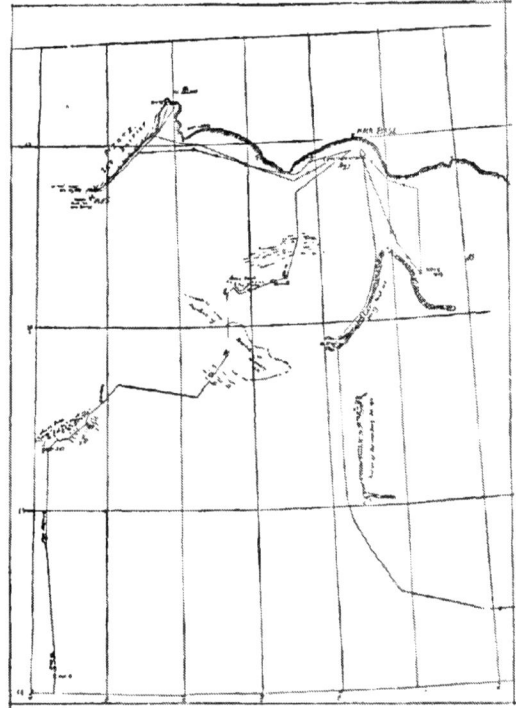

Tracks of the "Aurora" in the d'Urville Sea.

or no wind out at sea. It has been suggested that the comparative calm (felt a few miles from the coast) may only affect the surface, and that the violent wind still rages at a higher altitude.

The intensity of the wind (as recorded at the Main Base) appears to be associated with the abrupt contact of the cold air coming from the ice-covered land, and the comparatively warm air over the surface of the sea This would give rise to a difference of from 50° to 60° F. in the temperature of air over adjacent areas. Such a *difference* must cause very powerful circulatory movements.

The charts of Adélie Land, prepared under the direction of d'Urville, are very good, and, so far as I can form an opinion, remarkably accurate. He was favoured by fine weather on the day of his arrival, and his scientific observers were able to take observations at leisure. His judgment was sound, and his conclusions based on common sense

We must bear in mind that his brief " dash for the Pole " was a different affair from the organized exploring expedition under the command of Wilkes When d'Urville found that his special object could not be attained, he returned to Hobart ; and he could not have acted otherwise, as his crews were sickly, and his ships were not equipped for a long voyage in high latitudes.

The d'Urville Sea and Termination Ice-Tongue, so named by Dr. Mawson, commemorate the deeds of the gallant pioneers who led the way round the Australian Quadrant

The surface current along the northern edge of the pack-ice, so far as we could ascertain, sets to the westward The heavy swell, varying from north-east to northerly, which we often felt when following the northern border of the ice, has a tendency to balance any action of the wind in driving the looser edge to the north-west

At Termination Ice-Tongue, the current sets to the north-west,* being probably diverted by this obstruction. In my opinion this accounts for the enormous number of bergs, which, having floated along the edge of coastal pack, are set drifting round the tongue to the north-west, and are probably delayed near the end of the Tongue by various eddying currents.

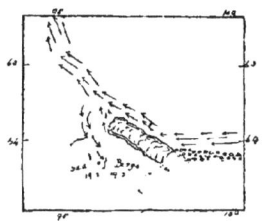

Surface currents off Termination Ice-Tongue.

* A north-west wind blows from the north-west A north-westerly current flows towards the north-west

THE D'URVILLE SEA

The collection of bergs met with to the west of the Tongue in 1912 and 1913 is carried round by the swirl of water off the end of the Tongue.

Vast numbers of bergs were sighted close to the 95th meridian. As we steamed north in 1913 they were still numerous in latitude 60° S.

To the south-west of Termination Ice-Tongue, a large area of open water—the Davis Sea—is due to the stemming of the pack-ice by the Tongue and the Shackleton Ice-Shelf which, in 1914, were found to be continuous.

It is very remarkable that Termination Ice-Tongue has not been broken off through pressure exerted by the pack-ice moving from the east during the winter. In 1914 we found ample evidence of the extent of this pressure. In my opinion, the Tongue must be partly resting on *land* to maintain its position. But, from the soundings obtained, there is no doubt that its Northern end is afloat.

The varying conditions of the ice at the western end of the field of operations of the Australasian Expedition are shown by the sketch maps for 1912, 1913 and 1914.

The party under Mr. Wild traced the coast-line of Queen Mary Land, from Gaussberg eastward to the great Denman Glacier, and discovered several islands, all ice-capped, rising through the Shackleton Shelf.

The shoal water just south of Termination Ice-Tongue would suggest that the northern part of the Shelf rests on land (possibly on rock), or on the debris left by the ice as it breaks off from the surface of the Shelf.

The examination of the edge of the Shackleton Ice-Shelf which was made in 1914, showed that on the western side it extended up to the southern end of Termination Ice-Tongue.

CHAPTER XXIII

THE "VINCENNES" IN 1840—THE "AURORA" IN 1912

Track followed by the "Aurora" in 1912, from Adelie Land westward, compared with the Track of the "Vincennes" in the same direction, as shown on Wilkes's Chart

The voyage of the "Vincennes," commanded by Lieutenant Wilkes, was, and will always remain, a remarkable achievement in the annals of Antarctic exploration. He was quite aware that the ordinary cruising vessels were not suitable for a voyage in high latitudes along an uncharted coast, and that the equipment of his ships left much to be desired. But he had been ordered to go by the Government of the United States, and this great sailor obeyed without hesitation. He traced out the icy barrier attached to Antarctica from as far as latitude 64° 1′ S., longitude 97° 50′ E., reporting the existence of high land, or the appearance of such, to the south of the line of impenetrable ice at several places along the track to the westward.

He was unfortunate in the matter of weather. Through fogs and snow, mid ice islands and bergs, he handled his ship during heavy gales in a way that thrills the heart of every sailor of the present day as he reads the story, written in unaffected style, but bearing the impress of *truth*.

I propose in this chapter to compare the track of the "Aurora" in 1912 from Adelie Land westward, with the track of the "Vincennes" in the same direction. There is no question but that certain errors of observation have been perpetuated in the official chart of the expedition of 1840. The latest issue of Wilkes's chart, with corrections, bears the date, August, 1914—

THE "VINCENNES" IN 1840—THE "AURORA" IN 1912

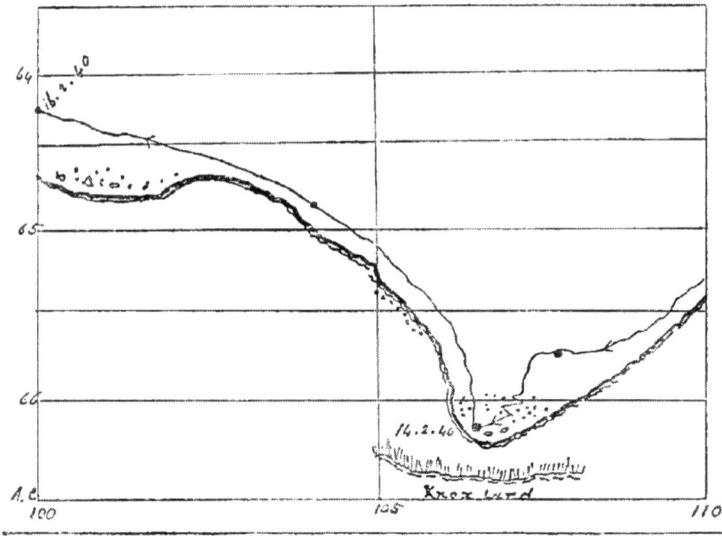

Wilkes's Chart : Knox Land (1840).

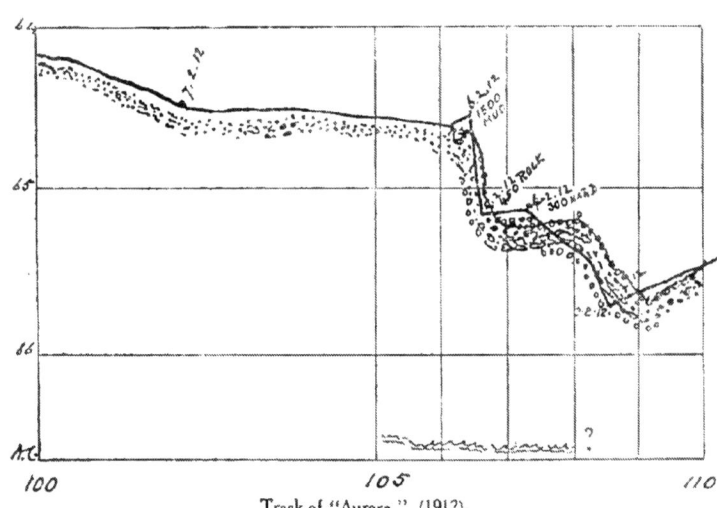

Track of "Aurora" (1912).

Track of the "Vincennes," 1840 and the "Aurora," 1912.

The first diagram shows the position of that part of the "Icy Barrier" named Côte Clarie by d'Urville. The high land just south of the barrier is an error. Wilkes was driven north from Piner's Bay by a violent gale, and, in my opinion, he traced this line as the *probable* direction of the coast-line to the west of Piner's Bay

The western point of the barrier, off which the "Vincennes" was hove to during the night of February 7th, 1840, was evidently the "Cape Carr" of Wilkes's log. With reference to "North's High Land," it may be a continuation of "Wilkes Land" (1912). A sounding taken on the "Aurora," close to the position marked "25.1.12." gave a depth of 230 fathoms on mud and rock, which is suggestive of land not far to the south.

The second diagram shows the position assigned by Wilkes to Totten's and Budd's "High Land."

On 31.1 12. the "Aurora" was in the position shown, with a sounding of 340 fathoms on mud. Totten's "High Land" should have been visible to the southward, as the weather was fine and clear, but nothing was visible over the heavy pack except a faint blue line which might have been a lead, but bore no resemblance to *high* land.

The line of pack ice trended to the north, and we were not able to confirm the position assigned to Budd's "High Land."

The third diagram shows the position assigned to "Knox Land" on Wilkes's Chart.

The "Aurora" made three attempts to push south through the heavy pack, but on each occasion was obliged to retreat owing to heavy pressure-floes. After the third attempt the ship was extricated with great difficulty, so we decided to continue westward towards the position marked on Wilkes's chart as Termination Land.

Wilkes had reported this as an *appearance* of land, seen to the westward over the "Icy Barrier." The "Gauss" had sailed close to the position in 1912, and H.M.S. "Challenger" had failed to reach it, by 15 miles, in 1874.

In justice to Wilkes, it should be remembered that he only reported an "appearance" of land.

There is no land in the direction indicated, but the ice-tongue seen by the "Aurora" in 1912 has been named

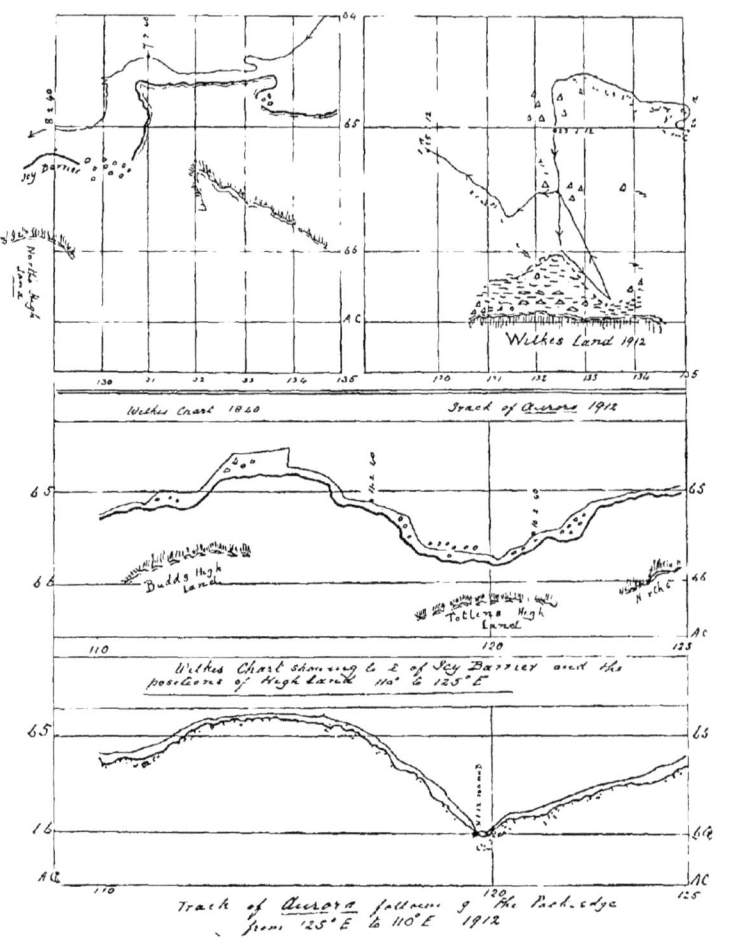

Diagrams to illustrate notes on Wilkes's Chart

Termination Ice-Tongue, in honour of the pioneer.

The diagram shows the position of "Termination Land" as seen from the "Vincennes" on 17.2.40, and as it still appears on the chart of the Expedition (latest issue).

The "Aurora" followed the edge of the pack, and, in the forenoon of 8.2.12., a long ice-tongue was sighted running in a north-westerly direction.

The position of the ship by D.R.* at noon is shown on the lower sketch.

Compared with the position of the "Vincennes" in 17.2.40., the American ship was in the same latitude—about eleven miles to the westward.

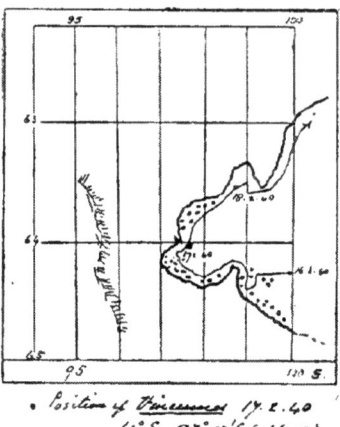

I am here taking the "Aurora's" position by D.R. at noon on 8.2.12. as correct. It is possible that it may have been in error, and could have been eleven miles too far to the east. The probability is that the position of both vessels—as far as longitude is concerned—is capable of such rectification, as the present position of Termination Barrier would make necessary, in order to account in an entirely satisfactory manner for Wilkes's report, which states: "Appearances of land were also seen to the south-west, and its trending seemed to be to the northward."

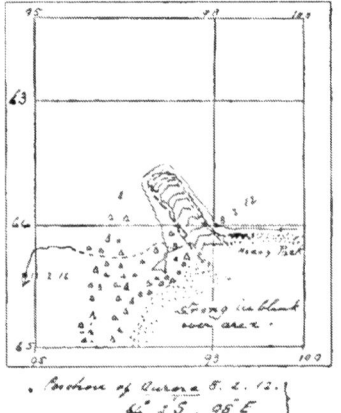

The "Aurora" rounded the Ice-Tongue, and, working

* D.R., dead reckoning, an estimated position unconfirmed by celestial observations, owing to cloudy weather or other reasons.

first through a berg-laden sea and later through open water, reached a point seventeen miles north of the actual coast-line, now called Queen Mary Land and some 180 miles to the south of Wilkes's position on February 17th, 1840.

Wilkes, animated by the true spirit of the explorer, determined to " carry on " in the face of great difficulties. When we remember that the " Vincennes " was a sailing ship, the way in which he managed to lay down the icy barrier attached to Antarctica, from Adelie Land westward, is highly commendable.

If we remove from his chart the " High Land " shown to the south of Côte Clarie and " Termination Land," the other positions of " High Land " may yet be proved to be approximately correct. Small errors in judging distances, or in observations for longitude, under the adverse circumstances of his voyage, must be allowed for in a generous spirit. In any case, some of the positions he gave have not been *disproved*.

The Admiralty chart of Repulse Bay, after Wilkes, was so largely responsible for my decision to continue west from the longitude of Knox Land that it may be mentioned.

I had freqently speculated on our way west, when looking over the chart, as to the cause of the peculiar northerly trend of the pack in the vicinity of Repulse Bay as shown on our charts. The only solution that seems possible is that either land or an ice-barrier interrupted the westerly drift of the ice at this point.

Although " Termination Land," which appears on Wilkes's chart had not been found by H.M.S. " Challenger " or by the " Gauss," I was convinced that an obstruction of some nature did exist here. I proceeded west to test the matter, and was not at all surprised when on the morning of February 8th, 1912, an ice barrier was reported not far from the position of Wilkes's " Termination Land."

In my opinion Termination Ice-Tongue not only existed in 1840 but was the cause of the arrangement of the ice around " Repulse Bay."

The similarity of the actual position of Termination Ice-Tongue and the appearance of the land reported by Wilkes are too striking to be a coincidence. Wilkes' chronometers need only have been a little out on the morning of the

17.2.40. to have placed him in a position when an "appearance of land to south-west trending northerly" would exactly describe "Termination Ice-Tongue" to any one some miles off who has not learnt that such huge masses of ice are not necessarily on a solid foundation.

It is well known that low clouds hanging over ice surfaces on a dull day bear an extraordinary resemblance to distant land. The blink from Termination Ice-Tongue would be brighter and harder than that of the pack surrounding him, and this was no doubt the reason why Wilkes recorded an "appearance of land" in this direction, although the actual Tongue itself might have been below the horizon or merged into the pack lying between it and the "Vincennes."

CLARIE LAND.—I have referred in Chapter XXII to the work of d'Urville, whose discovery of Adelie Land preceded the visit of Wilkes to Piner's Bay, by nine days.

Comparing the observations of d'Urville and Wilkes, there is little doubt that they both sighted and described the huge ice formation (which has since disappeared) named by d'Urville Côte Clarie.

As I have already stated Cape Carr from Wilkes would probably be the western extremity of this mass of ice.

CHAPTER XXIV

L'AVENIR

When the Motive is right and the Will is strong
There are no limits to human power.
—E. W. Wilcox

Although much work has been done in, and many volumes have been written about, Antarctica, a vast amount of work remains to be done. The ambitious explorer will find ample scope for his energies in the African Quadrant or in the region to the east of King Edward VII Land.

The crossing of the Continent from sea to sea is a most interesting problem, and there is little doubt that, sooner or later, others will try to penetrate that dangerous area called the Weddell Sea, to establish a base on the coast-line from which an advance might be made to cover the 1,500 miles across the Antarctic Continent. Such a land journey is a very big undertaking, but, with the improved equipment of recent years, is not, by any means, impossible.

It is very desirable that the work of laying down the actual coast-line of the Continent, begun by the Australasian Expedition, should be continued. The African Quadrant stretches for 90° westward from Gaussberg; a promising field for exploration. It is not too much to hope that South Africa will regard the Quadrant bearing its name as a coastal region to be explored, in which venture the inhabitants of the Union should take a leading part.

Oceanographical research should be developed by future expeditions in the Antarctic and Sub-Antarctic regions. The detailed investigation of certain localities, which have been "prospected" by pioneers, would in all probability lead to very interesting discoveries. By the help of the improved apparatus of the present day and with the necessary time, a great field of work is open in this direction.

The next Antarctic expedition should have a vessel

constructed for the work. The introduction of oil fuel, the varied necessities of a modern scientific expedition, the desirability of carrying the means of aerial reconnaissance, all call for a specially designed vessel. A steel-built ship, reinforced with wooden beams, like the "Beothic," would probably cost less than a vessel of the type of the "Discovery."

The necessity of special training for Antarctic work is not essential, except in the case of the leader. The leader selects the members, each being a specialist, well versed in his own subjects. A knowledge of navigation and elementary surveying is useful to every explorer. The man of thirty is probably at his best, but, from my own experience, I am inclined to believe that from twenty to twenty-five you get the real enthusiast whom nothing daunts, and who is never tired. The Australians proved born explorers, though without polar experience. They were always keen, always resourceful, and never cast down by difficulties. Camping privations and bush life in the less settled parts of Australasia may, to some extent, account for this.

Recent expeditions have had to beg for funds. Really useful work has too often been sacrificed to the purely spectacular. The explorer, who is handicapped by debt, may be tempted to stimulate the public with sensational feats : the temptation is difficult to resist—or justify.

To the explorer who has not the money to provide good equipment of every kind, my advice is—" Keep out of the Antarctic ! "

To those who intend to follow the lead of the pioneers who, during the past hundred years (1819–1919), have sailed through the Southern Ocean, or made journeys over the inland ice, I should say—" Study the literature of the Antarctic, make yourself thoroughly acquainted with the nature of the obstacles and the difficulties to be encountered in those regions." I can realize the immense value of the information on these matters which I obtained from books. This is my apology for penning these lines.—*Vale.*

Appendix I

LIST OF MEMBERS OF THE AUSTRALASIAN ANTARCTIC EXPEDITION

DOUGLAS MAWSON, D Sc., B E , Commander of the Expedition

The Staff at the Main Antarctic Base, Commonwealth Bay, Adelie Land, included :—

[1] Lieut. R. Bage, R.A E	Astronomy, Magnetics, and Tides
C. T. Madigan	Meteorology
[2] Lieut B E. S. Ninnis, R.F.	In charge of dogs.
[3] Dr. X Mertz	In charge of dogs
Dr A L. McLean	Chief Medical Officer, Bacteriologist
F H. Bickerton	Air-tractor sledge
A. J. Hodgeman	Cartographer and Sketch Artist
J. F. Hurley	Official Photographer.
E N Webb	Chief Magnetician.
P E Correll	Mechanician and Assistant Physicist.
J. G. Hunter	Biology
C F Laseron	Taxidermist and Collector
F. L. Stillwell	Geology.
H D Murphy	Stores
W H. Hannam	Wireless Telegraphy
J. H. Close	Assistant Collector
Dr. L. A. Whetter	Surgeon.

The Staff of the Western Antarctic Base comprised :—

F Wild	Leader.
A. D. Watson	Geology.
Dr S E Jones	Medical Officer
[4] C T. Harrison	Biology.
M. H. Moyes	Meteorology.
A. L Kennedy	Magnetics
C. A Hoadley	Geology.
G. Dovers	Cartography.

[1] Killed in action at Gallipoli, 1915
[2] Lost in a crevasse, King George V Land, 1912
[3] Died of starvation, King George V. Land, 1913
[4] Lost on the " Endeavour " in the Southern Ocean, 1914

APPENDIX I

The Macquarie Island party comprised :—

G. F. AINSWORTH	Meteorology, also in charge of the station.
[1] L. R BLAKE	Geology and Cartography.
H. HAMILTON	Biology
C A SANDELL	Wireless Telegraphy.
A. J SAWYER	Wireless Telegraphy

The ship's company numbered twenty-five, including :—

J K. DAVIS	Master of the *Aurora*.
N C. TOUCHER	Chief Officer, first Antarctic Voyage.
F D. FLETCHER	Chief Officer, second Antarctic Voyage and during the Spring and Winter Sub-Antarctic Cruises.
J H BLAIR	Chief Officer, final Antarctic Voyage
P. GRAY	Second and Navigating Officer.
C P DE LA MOTTE	Third Officer.
F J. GILLIES	Chief Engineer.
H CORNER	Second Engineer.
J FORBES	Sailmaker
G. WILLIAMS	Chief Steward.
J RUST	Cook.
O. McNEICE	A B.
J McGRATH	A.B.
H. HACKWORTH	A.B.
L. VINJI	A.B

The following six gentlemen accompanied the vessel on one or other of the voyages :—

SIDNEY JEFFRYES	Wireless operator, who relieved W. H. Hannam during the second year.
E R. WAITE	(Curator Canterbury Museum, Christchurch), Biologist, first sub-Antarctic cruise.
Professor T. FLYNN	(Hobart University), Biologist, second sub-Antarctic cruise.
J VAN WATERSCHOOT VAN DER GRACHT	Landscape artist, second Antarctic cruise.
Captain JAMES DAVIS	Whaling authority, second Antarctic cruise
C. C. EITEL	Secretary, second Antarctic cruise.

[1] Killed in action in France, 1918

Appendix II

BRIEF NOTES ON SOME ICE-FORMATIONS IN THE ANTARCTIC REGIONS

The following notes are intended to illustrate some of the terms used in the narrative with reference to ice I have tried to explain the difference between land-ice and sea-ice by referring to the origin of each, and to trace the stages through which "one-year-ice" passes in the course of twelve months

No words of mine can convey a true idea of the massive proportions, the vivid colours or the marvellous beauty of a group of Antarctic bergs, seen on a bright day. Nor can I give an adequate description of the irresistible *power* of floe-ice, when driven by wind and wave, so that the notes must be taken as roughly explanatory of the terms used in connection with the different kinds of ice to be met with in the Antarctic.

1. The "Great South Land" is surrounded by the Southern Ocean, and, over the area covered by that Ocean and within the average limit of the floating-ice, there are, even in summer, wintry conditions; so much so that a blizzard off the land may cause the temperature to fall below zero Fahrenheit **LOW TEMPERATURE**

In winter, temperatures as low as 60 degrees of frost ($-28°$ F), accompanied by hurricane winds, were recorded at Commonwealth Bay.

2 The ice met with in polar regions is of two kinds— land or fresh water-ice and sea-ice, the freezing point of fresh water is $32°$ F., but sea-water begins to freeze at about $29°$ F. Land-ice three inches thick, will bear the weight of a man, but sea-ice of the same thickness could be easily penetrated by a stick **LAND-ICE SEA-ICE**

3. The continent of Antarctica is covered with an ice-cap rising to a height of 10,000 feet near the South Pole, and the accumulation of snow, year after year, causes a continuous movement from the centre towards the coast. Generally the edge of this glacial cap takes the form of an ice-cliff broken here and there by glacier **ICEBERGS**

tongues. As the cliff is pushed seaward by the overwhelming force behind, masses of ice are broken off ranging from half a mile to 40 miles in length These huge masses are formed of glacier-ice, that is to say, they have been built up from a base of glacial-ice by successive snow-falls. Smaller bergs are formed when irregular pieces of ice break away from the snouts of glacier tongues. As bergs drift into lower latitudes, they get undermined by the warmer waters and break up into smaller bergs and irregular pieces of intense hardness, known as "growlers."

SHELF-ICE 4 The vast field of ice to which Dr. Mawson gave the name of the "Shackleton Ice-Shelf" extends to a distance of about 180 miles from the coast line of Queen Mary Land, and covers an area of several thousands of square miles. It originated from the glacial flow over the plateau to the South, while, every year, an additional layer of consolidated snow has been added to its surface by the frequent blizzards These additions are clearly marked on the white face of the ice-cliff. The Great Ross Barrier is of similar formation. Bergs breaking away from shelf-ice are generally flat-topped with nearly perpendicular sides and from 150 to 200 feet in height.

SEA-ICE

BAY-ICE 5. Sea-water begins to freeze at about 29° F., but a fall of ten degrees in temperature will cause the layer of thin ice to increase rapidly in thickness. Any new ice, from one to six inches thick, is called bay-ice, as it forms more easily in sheltered positions. At first this ice is dark in colour and *gluey* in texture, but when it has increased in thickness under the action of frost, it becomes whiter and more brittle

PANCAKE-ICE 6. Any disturbance of the surface of the sea will cause this crust of bay-ice to break up into cakes, more or less regular in shape; then the lines of fracture come together again, and the sides and corners of the cakes get crushed and broken and the edges become turned up This process goes on while the cakes are becoming thicker and more compact, that is, if the temperature falls lower. Such ice is known as pancake-ice.

FLOE-ICE 7. As the winter frosts continue the pancakes get frozen together, and, eventually, form a sheet of ice, from five to six feet thick, which are only broken

APPENDIX II

up by violent weather conditions. Such sheets are known as floe-ice. A large piece of ice which has been detached from an ice-sheet may be called a "floe," but this seldom occurs in new ice, that is, in ice formed during one season

When the surface of the ice is unbroken as seen from the crow's nest of a ship, the sheet is called field-ice FIELD-ICE

8 When the surface of the sea is violently disturbed during a gale, the movement of the water beneath the floe-ice causes the latter to break up into pieces of various sizes. Driven by the wind and tossed about in a tempestuous sea the broken-up floe becomes a chaotic mass, which may include the smaller bergs and the debris of bergs This is known as pack-ice. The usual time of the break-up is in spring, when the rising temperature is causing the sea-ice to become rotten PACK-ICE

In the course of the break-up, the floes are crushed together and their edges are broken and piled up by *pressure*. If the break-up occurs during the winter, the whole mass freezes again almost immediately, and the internal stress and strain becomes intense, resulting in the formation of *pressure-ridges* of broken ice. The piles of ice formed by pieces of the "pack" are known as *hummocks* PRESSURE RIDGES

HUMMOCKS

9 "Brash-Ice" is often visible on the edge of the pack; it is composed of small pieces which have been broken off bergs or floes during their struggle with wind and sea and with one another. BRASH-ICE

10. The pack drifts about under the combined influence of winds and currents In the Ross Sea area, a large portion of the pack drifts into warmer waters and melts during the summer. Between Adelie Land and Termination Ice-Tongue, the pack forms a belt or fender along the coast, outside the fast sea-ice, but the conditions vary considerably from year to year. (See Chap XXI)

11. When a ship is working through the pack, the proximity of open water is shown by a dark line on the horizon known as a water-sky. WATER SKY

12 When approaching a belt of pack or any large formation of ice, a bright reflection above the horizon indicates that the *ice* is not far distant. This is known as an ice-blink ICE-BLINK

13 Floating masses of pack-ice may be either "loose"

APPENDIX II

or "close." A vessel constructed for navigation in polar waters can work through the former without much trouble, but, if the pack is "close," it is sometimes advisable to coast along the edge and wait for better conditions

BESET — Pushing into heavy pack may result in the ship getting *beset*—which means that, after it has worked through the obstacle for a certain distance, the ice has closed round the vessel and rendered further movement impossible. A vessel becoming beset late in the season will, in all probability, not get free before the following summer.

The *Gauss* was beset and was released only after a detention of 12 months. The *Endurance* was beset in the Weddell Sea, and was crushed by the ice after drifting for several months

ONE-YEAR-ICE — 14. Field-ice and floe-ice formed during a single winter are known during the following summer, as "one-year-ice"

OLD-ICE — At the beginning of the second winter the floes may average seven feet in thickness. The severe frosts and heavy snow-falls of a second winter will add considerably to the thickness, and, when this ice breaks up in the second spring, the pack-ice resulting is a more serious obstacle to navigation than the one-year-ice

ICE-STREAMS — When irregular pieces of all kinds—fragments of bergs and floes—are driven together by the action of wind and sea, they form *streams* of ice which nearly always lie at right angles to the wind. If a gale comes on, one stream joins another, and, as the storm increases, a formidable pack is forced onward until checked by meeting another solid pack or an ice-cliff. If this pack should be driven against the edge of a cliff like that at Termination Ice-Tongue, it will indent deeply the face of the cliff and fall back again upon itself in a heap of shattered fragments.

CONSOLIDATED PACK — Again a further change may occur—perhaps it is due either to wind or tide. The streams of ice which have been forced together open up, and small channels and lanes of calm water appear. Then with the advent of cold weather all this loose pack is frozen together, the lanes are covered with new ice, wind may pack the old ice and the new into solid floe, and many floes may finally unite to form great "*fields*" of ice—consolidated pack.

Appendix III

THE ANTARCTIC REGIONS AS KNOWN IN 1914

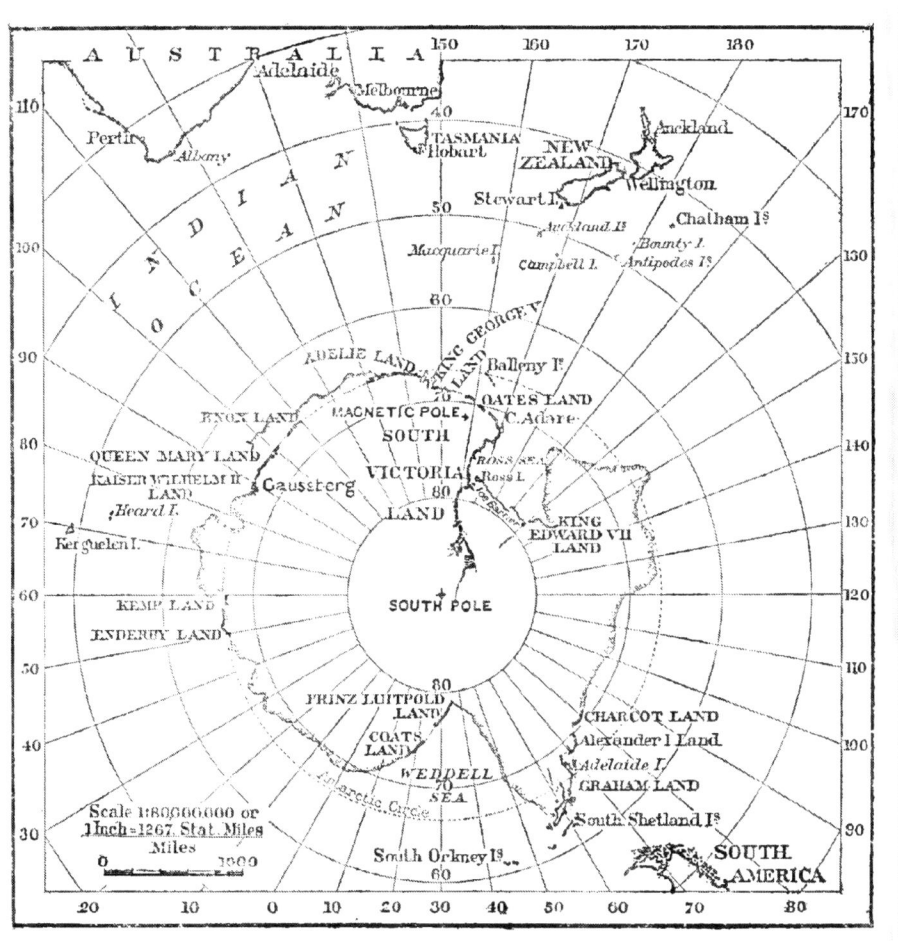

Published by the Royal Geographical Society.

Appendix IV

PLAN OF THE AURORA

Aurora.—Auxiliary barquentine. Tonnage, 580 gross, 380 nett Built 1876 by Stephens, of Dundee. Length, 165 feet. Breadth, 30 ft. 6 in Depth, 18 ft. 9 in. Loaded draught, carries approximately 600 tons, deadweight. Engines—steam-compound. Nominal horse-power, 98.

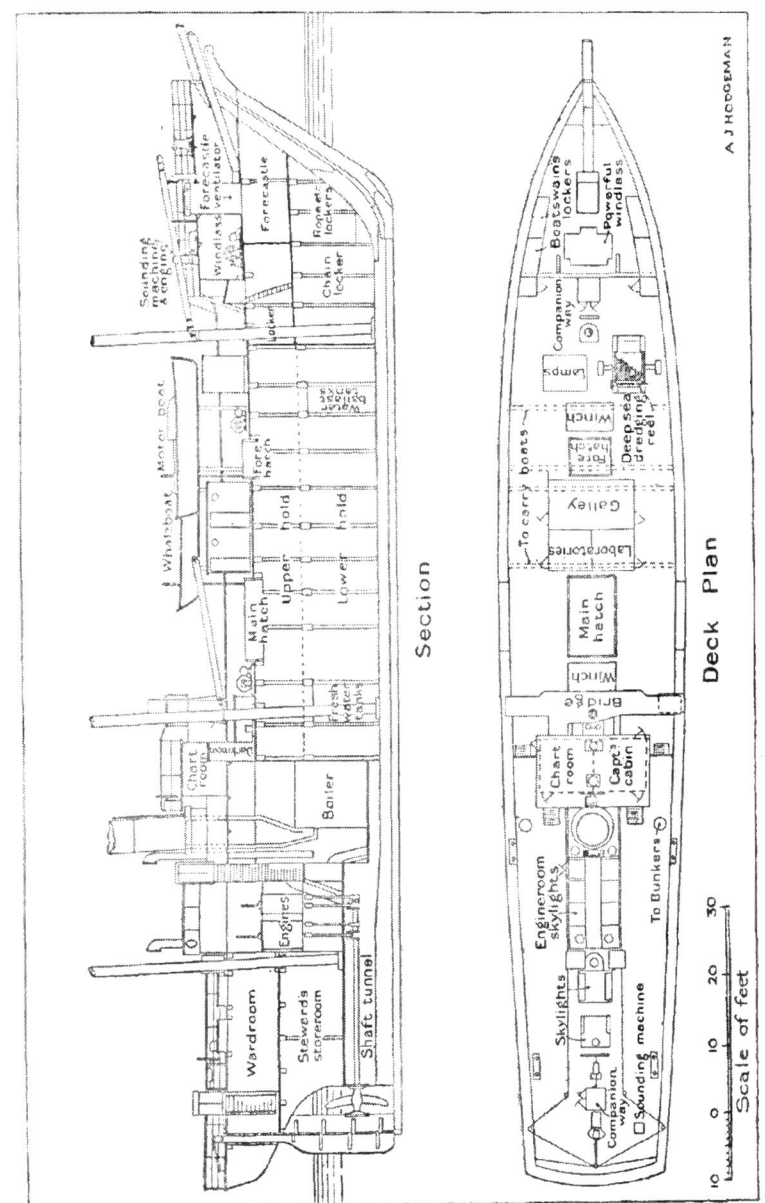

PLAN AND SECTION OF THE S.Y. *AURORA*

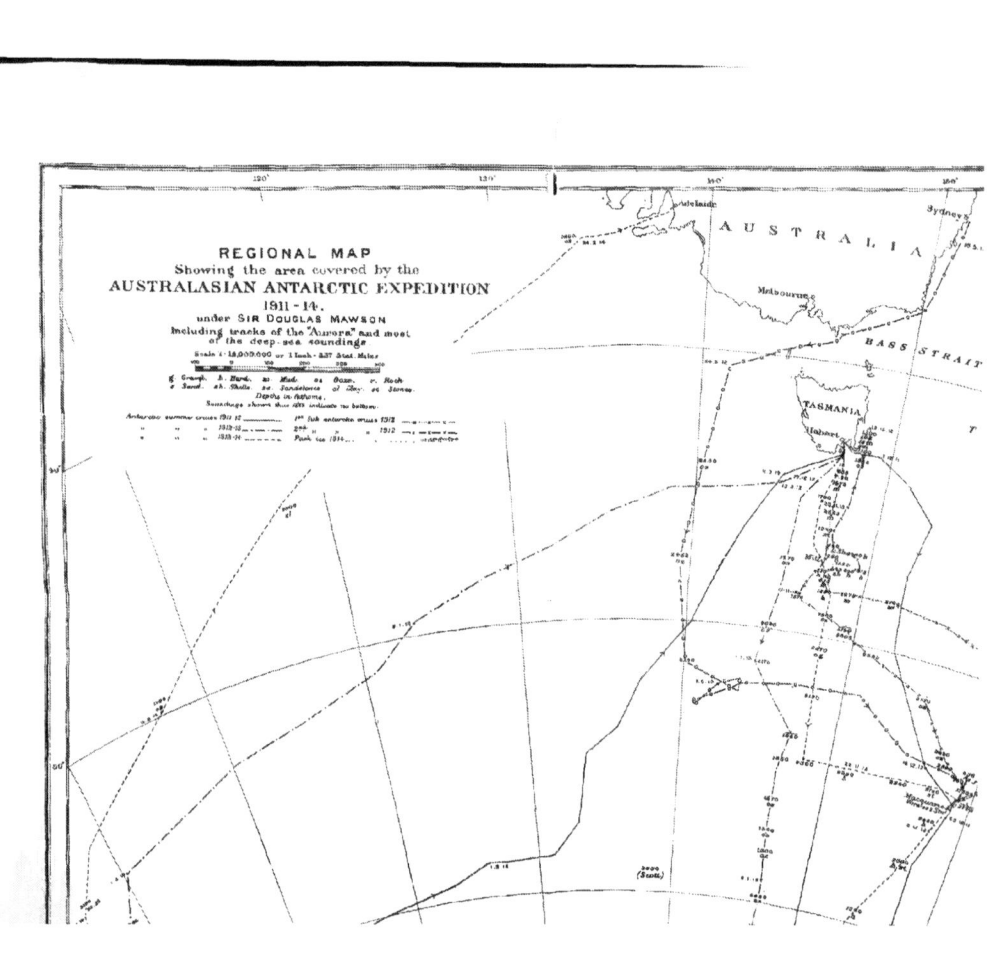

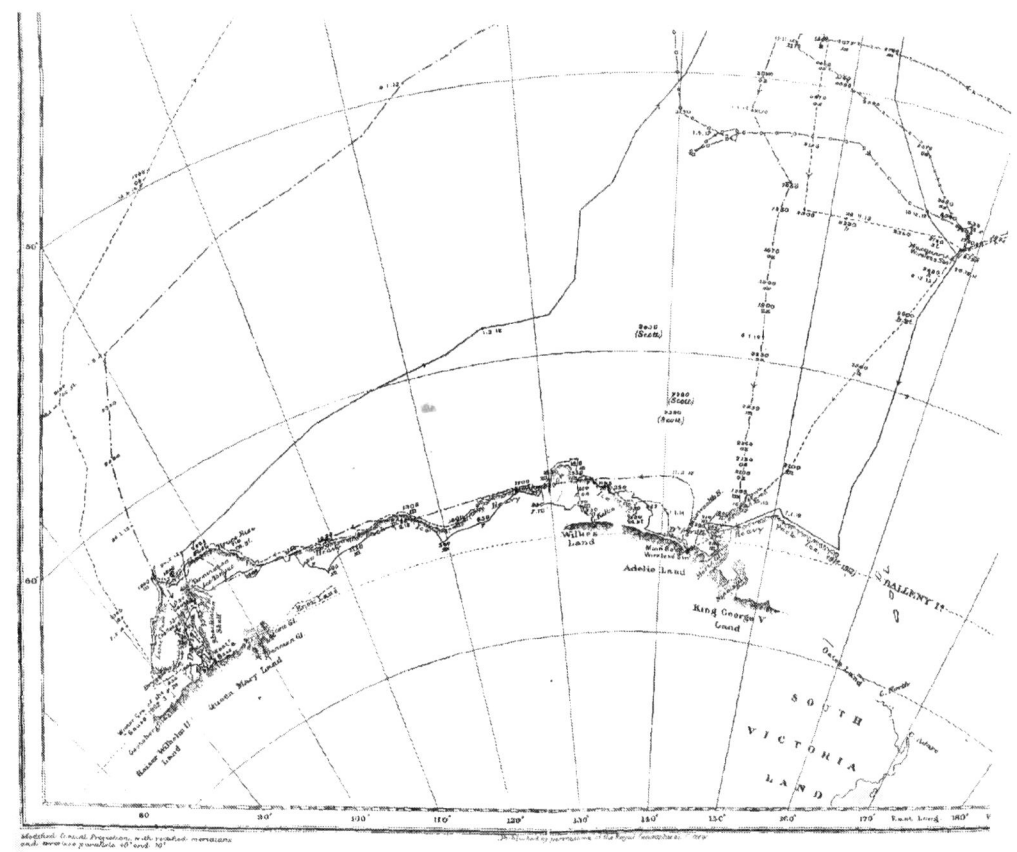

INDEX

Adelaide, return of expedition to, 149
Adelie Land, 164; charts of, 158; discovery of, 33, 35, 36, 44, 46, 64, first landing on 28; limits of pack ice near, 153, 154
Ainsworth, G. F., 20, 22, 64, 65, 109, 116, 121
"Aladdin's Cave," ice cave near Commonwealth Bay, 88, 91, 96
"Amukura," steamer, 71
Amundsen, Captain, 54, congratulation to, 55; dogs presented by 82
Antarctic Circle, first crossing of, 1, by "Aurora," 28
Antarctica, area of, 1, ice-formations in, 170–175; suggestions for exploration of, 166, 167
"Astrolabe," d'Urville's voyage in, 35, 44, 153
Auckland Islands, 69
"Aurora," Antarctic Research Ship, 2, 5, 8, 9, 10, 14, 49, 52, 137, 153, 154, 159, 162, 163, equipment of, 10, plan and dimensions of, 176; refitting of, 10, strikes submerged rock, 18
Australasian Antarctic Expedition, 2, land on Adelie Land, 28, main base of, 31, members of, 3, 168, 169, objects of, 2

Bage, Lieut. R., 28, 89
Balleny, Captain John, 34, 40, 47
Ballast, 54
Barrier, Ice, 25, 39 42, 45, 52, 53, 143, 144, 146, adrift, 99
Bass Strait, 57, 61, 72
Bauer, Mr., 19, 21
Bay Ice, 86; definition of, 172
"Beothic," 167

Beset, explanation of, 174
Bickerton, F H, 28, 89
"Bishop and Clerk," rocky islets, 18, 23
Blake, F H, 20, 64, 109, 116
Blanche Rock, 71
Blizzards, 26, 27, 49, 87, 88, 102
"Blue Billies," 84, 121
Brash Ice, 24, definition of, 173
Bristow, Captain, 69
Bruce, Dr W. S, 3, 11, 58, 136; rise, 136
Buchanan, J Y, F.R S, 32
— Bay, 131
"Budd's High Land," 39, 162
 [65
Canterbury Museum, specimens for,
Cape Carr, 38, 46, 162, 165
— Denison 31, 43
— Discovery, 44
— Frederick Henry, 16
— Hunter, 129
Capetown, Arrival at, 12
Carnley Harbour, 69, 70
Caroline Cove, 18, 21, 22
"Challenger," H M S, 32, 33, 40, 162
Christmas of 1911, 23, of 1912, 83, of 1913, 129
Close, J H, frost bitten, 88
Coal, 15
Commonwealth Bay, 31, 32, 43, 64, blizzard at, 87, 88, return to, 87, 124
Commonwealth Government, grant from, 108
Compass, disturbance of, 18, 43
Consolidated Pack, origin of, 174, 175
Cook, Captain, pioneer Antarctic navigator, 1, 39
Cook, Sir Joseph, G C M G, 108
Correll, P E, 89, 109

INDEX

Côte Clarie, 30, 34, 36, 45, 46, 154, 162, 164, 165, non-existence of, 85, 154
Currents, 105, 148, 158, ocean, 110

Danish Government supply sledge dogs, 11
Dannevig, H C., drowning of, 74
David, Professor T. W E, 2; message to, 64
Davis Captain J, of Hobart, 82, 85, 89
Davis Sea, 159
Denman, Lord, 108, 150, glacier, 159
"Discovery," 167
Dogs, Greenland sledge, 11, 16, 20, 31, 82
"Drake," H.M S., wireless from, 84
Drygalski, Professor von, 34, 41, 42; island, 139, 148, high land reported by visited, 104
d'Urville, Captain Dumont, 33, 35, 40, 44, 46, 154, 158, 162, 165
— Sea, 156, 158
Dundee, 5

Eitel, C C, 21, 82
"Eliza Scott," 33, 34
Elliott's Reef, 19
"Endeavour," fishery cruiser, 74, loss of, 74
Enderby Brothers, 34, 69
— Island, 69
"Erebus," 40
Erebus Cove, 71, 72
Evans, Commander, R.N., contribution from, 107

"Fantome," H M S, 16
Ferguson, Mr., 109
Field-Ice, definition of, 173
Fletcher, F D, chief officer of "Aurora," 89, 113
Floe Ice, 25, 46, 50, 53, 145, 159; definition of, 173; heavy, 102
Flynn, Prof T, 75
Fossilized wood, 130
"Fram," Antarctic research ship, 54
Freemantle, wireless messages to, 64

Gales, 14, 19, 29, 30, 31, 47, 54, 63, 64, 91, 94, 100, 126, 135, 141
"Gauss," Antarctic research ship, 33, 34, 42, 138, 162, 164, early besetment of, 98, 100
Gaussberg, 2, 3, 34, 42, 159
Gillies, F. J, chief engineer of "Aurora," 94
Gipps, Sir G, 40
Glacial mud, 26
Glacier tongues, 49, 50, 93, 122, 125, 130, 151, 153
Globigerina ooze, 62
Gracht, J van der, marine artist, 82
"Grafton," schooner, wreck of, 70
"Grasshopper," air-tractor sledge, 15; damage to, 17
Gray, P, second officer of "Aurora," 121
Greely, expedition of, 5, 6
Green Island, 71
Groom, Hon. L E, 150
Gulf of Mexico, 110
Gulf Stream, 110, 111

Hamilton, H, 20, 67, 109
Harrison, C T., drowning of, 74
Hasselborough Bay, 19, 24
Hatch, Mr, lessee of Macquarie Island, 19, 65
Henderson, Mr, 109
Henderson, Professor G C, 2
Hippon, Anthony, master mariner, 12
Hobart, 15, 35, 37, 54, 55, 75, 83, 105, 109
Hodgeman, A J, 89, 96
Holyman, Captain, 16, 21
Hummocks, origin of, 173
Hunt, H. A, Commonwealth meteorologist, 16; successful forecasts by, 16, 117
Hurley, J F, 28, 96, 109
Hurricane, 128

Icebergs, 24, 26, 30, 46, 85, 122, 143, 144; aground, 53; origin of, 171
— blink, 45, 85, 135, 145, 165, definition of, 173
— cave, 130, 131

INDEX

Ice, inland, 140
— island. *See* Icebergs
— limits, observations on, 151–154
— streams, origin of, 174

Jan Mayen, island of, 5
Jeffryes, Sidney, wireless operator, 82
"Jessie Nicholl," wrecked schooner, 19

Kaiser Wilhelm II Land, 41
Kelp, 110
Kelvin sounding machine, 59
Kennedy, A L, 28
King Edward VII Land, 166
King Island, 61
Knox Land, 39, 48, 162, 164
Kerguelen, 32, 40, 110

Laurie Cove, 71
Lloyd-George, Rt. Hon David, secures Imperial Government grant for Relief Expedition, 107
Lucas sounding machine, 57, 58, drum of, bursts, 62
Lucas-Tooth, Sir R, handsome donation from, 107
Lusitania Bay, 65, 119

Mackellar Islands, 86, 124, 127
Mclean, Dr A L, 89, 96
Macquarie Island, 57, 62, 63, 113, description of, 18, 19; map of, 116, observers at, 20; ocean depths near, 118; weather at, 116
— — *Daily News*, 64
Madeira, Island of, sighted, 11
Madigan, C T, 28, 151, frost bitten, 88, report from, 95
Magnetic pole, 35
— station, 31
Masson, Professor Orme, 2, 150
Mawson, Dr, 2, 16, 21, 24, 28, 30, 31, 32, 64, 89, 98, 124, 127, 143, 147; non-return of, 90, 93, search party for, 93; wireless message from, 97
— Relief Fund, 107
Mertz, Dr. X., 3, 11, 15, 89; death of, 97, 106

Mertz, Glacier Tongue, 93, 122, 125, 130, 151, 153
Meteorological stations, 64
Mill, Dr H. R., conversation with, 107
— Rise, Oceanic ridge, 80
Monaco, Prince of, 3
"Monagasque trawl," 59
Murray, Sir John, K C B , 107
Musgrave, Captain, 70

Nansen, Dr , eulogy of Captain Scott, 105
Needle Rock, 15
News Agency Station in N S W., wireless messages from, 64
New Year of 1912, 25
Ninnis, Lieut. B. E. S , 11, 15, 20, 89; death of, 97, 106.
— Glacier-tongue, 151
North-East Bay, 78
"North's High Land," 162

Observation Point, 72
Octopus, 80
Ocean, serial temperatures of, 110
"Ocean " ship, 69
Old-Ice, definition of, 174
One-Year-Ice, definition of, 174

Pack ice, 24, 25, 44, 45, 46, 47, 48, 49, 85, 99, 100, 122, 131, 133, 134, 135, 140, 147, 148, 149; definition of, 173
Pancake-ice, definition of, 172
"Peacock," Antarctic research ship, 37
Peake, Hon. Mr., 150
Piner's Bay, 33, 37, 44, 162, 165
Penguins, 113
— Adelie, 28
— Emperor, 50, 102, 141
— new variety of, 73
Petrels, 73, 100
— Antarctic, 24, 123
— Cape pigeons, 100
— giant, 25
— Prions, 84, 121
— silver grey, 131; eggs of, found, 131

Petrels, snow, 123
"Porpoise," Antarctic research ship, 37
Port Jackson, 37, 57
— Lyttleton, 73
— Ross, 69, 71
"Post Office Stones" at Table Bay, 12
Power, Mr., 109
Primmer, Mr., 57

Queen Mary Land, 151, 159, 164

Rabbits, 73
Reefs, 43
Reid, Sir George, 107
Repulse Bay, 164
Rock Gripper, 109
Royal Company Islands, 58, 61, 62, 64
Royal Geographical Society, donation from, 2
Rookery Island, 141
Ross, Sir James Clark, 35, 40, 72
— Barrier, 3, 53

Sabrina Land, 33, 35, 48
— loss of cutter, 34
Sandell, C. A., 20, 64, 109
Sawyer, A. J., 20, 64
"Scotia," Antarctic research ship, 58
Scott, Captain Robert Falcon, 3; news of death received, 105
— Lady, assistance from, 107
Scottish National Antarctic Expedition, 58
Sea bottom, deposits, 26
Sea Lions, 70
Seals, 145
— Ross, 141
— sea-elephants, 65, 66
— sea-leopard, 24
— Weddell, 26, 50
"Sea of Bergs," 103
"Shack," the, 64, 119
Shackleton ice shelf, 53, 102, 104, 142, 143, 154, 159
Shelf-Ice, origin of, 172
Sounding, Lucas machine for, 10

Soundings, 18, 26, 27, 28, 44, 46, 48, 49, 50, 62, 66, 73, 76, 78, 79, 80, 83, 84, 85, 86, 109, 110, 113, 121, 122, 126, 128, 130, 131, 133, 134, 136, 137, 139, 141, 142, 144, 146, 147, 149, 154, 155
South Island, 69
Squalls, 21, 31, 86, 94, 129
Storm Bay, 17
St. John's, 5
Sydney, 15, 37, 40, 57
Suva (Fiji), wireless messages to, 64
Swell, 25, 45, 52, 63, 119, 149

Table Bay, "Post Office Stones" at 12
Tasman, sights New Zealand, 1
Termination ice-tongue, 54, 100, 102, 146, 147, 151, 153, 155, 158, 159, 163, 164, 165, 174
— Land, 39, 49, 100, 162, 164
"Terra Nova," 24, 105, 121
"Terror," 40
— Cove, 72
"The Shack," 64, 119
"Toroa," S.S., 16, 21
"Totten's High Land," 48, 162
Trawling, 76, 80, 128, 130, 131, 133, 134, 136, 139, 142, 145
— equipment, 59, 60

Victorian Government, liberality of, 108
"Vincennes," Antarctic research ship, 27, 33, 37, 38, 40, 44, 163
— track of, compared with "Aurora," 160, 165

Waite, E. R., 57, 62, 65, 66, 72
Water sky, definition of, 173
Way, Sir Samuel, 150
Weather Bureaux, Dominion, 64; Federal, 64
— reports, 64
Webb, E. L., 31
Weddell Sea, 166
Western party, measures for relief of, 98

INDEX

Whales, 85, 100, 135
— finner, 137
— killer, 141
Whetter, Dr L. A, frost bitten, 88, 89
Wild, Frank, 3, 28, 30, 32, 48, 50, 52, 53, 159, relief of, 102, 103, western base of, reported, 64

"Wilkes Land," 42, 153, 162
Wilkes, Lieut, 27, 33, 37, 40, 44, 46, 48, 158, 160, 162, 163, 164, 165
Wireless, 11, 20, 89
— weather reports, 64

"Zélée" corvette, 35

Printed in Great Britain by Butler & Tanner, *Frome and London*

Reprinted from
INDUSTRIAL AUSTRALIAN
and Mining Standard
January 8, 1920

CURRENT LITERATURE

The Wild Antarctic Seas.—Three different methods of exploration have been employed to wrest her secrets from the icebound Southern Continent. Two of these dealing with the Polar Plateau and with land journeys along the coast have been described in many books well known to the Australian public. It has remained for Captain John King Davis to tell us of the third in his recent book "With the Aurora in the Antarctic." Here we find the story of his unrivalled oceanographic work in the waters between Australia and Antarctica.

Although in other expeditions valuable results were gained by the exploring vessels during their cruises in the winter months, none of these equals the work done by the Aurora, and none of the masters has hitherto given us more than a summary of his investigations. The present volume is of special interest to Australia for many reasons, and not least because it deals with one of the most important of the fields of work of the Australasian Antarctic expedition. This is emphasised by the leader in his introduction to the book. Sir Douglas Mawson states: "I counted upon the co-operation of John King Davis as second in command, to take charge of the ship. In him I had every trust and confidence. . . . Davis and his men have carried out a piece of research which in scientific value is comparable with that accomplished ashore."

An early chapter deals with Macquarie Island, that queer ridge of glaciated rock which may be described as a tile set on edge athwart the wild west winds. All round this dependency of the Commonwealth the Aurora made soundings which revealed the profundity of the depths surrounding it. Then we are told of the mysterious Royal Company Isles, which were "discovered" in 1776 by a Spanish captain about 400 miles south of Tasmania. Unfortunately for the Spaniard Davis sailed right across the alleged position of these islands, which must now disappear from the charts. Another interesting chapter deals with the Auckland Isles to the south of New Zealand. We should have liked more particulars of the castaways and settlements which have enlivened the history of these desolate sub-antarctic islands.

In November, 1912, Davis was dredging 250 miles south of Tasmania when his apparatus was carried away by what may best be described as a crag of a 'drowned Tasmania.' Rising 8000ft. above the ocean floor he found a large plateau—"Mill Rise"—which raises all sorts of questions as to a former connection between Australia and Antarctica.

The most vital chapters in the book deal with his hazardous voyages along the icebound boast of Antarctica. The reviewer has spent many months sledging in the Antarctica and has had some slight experience of aviation, but for supreme danger and discomfort he places easily first a cruise among the bergs in the twilight of a polar autumn. Yet this was a commonplace to Captain Davis—the most experienced navigator of Antarctic waters.

We read much of the treachery of the famous blizzards. For instance, while wait-

described as a crag of a 'drowned Tasmania.' Rising 8000ft. above the ocean floor he found a large plateau—"Mill Rise"—which raises all sorts of questions as to a former connection between Australia and Antarctica.

The most vital chapters in the book deal with his hazardous voyages along the icebound coast of Antarctica. The reviewer has spent many months sledging in the Antarctica and has had some slight experience of aviation, but for supreme danger and discomfort he places easily first a cruise among the bergs in the twilight of a polar autumn. Yet this was a commonplace to Captain Davis—the most experienced navigator of Antarctic waters.

We read much of the treachery of the famous blizzards. For instance, while waiting to take off Mawson's party at the main base, the Aurora was anchored a short distance from the shore. The boats had just taken advantage of a period of calm to land some stores. Suddenly a single terrific gust struck the ship, snapped the anchor chain and blew the Aurora far to the north.

A week later Davis had one of his most perilous voyages while racing the winter darkness to rescue Wild's party far to the west.

Picture them sailing through a sea of bergs scattered over a belt 300 miles wide. Yet Davis calmly dismisses this thrilling experience with the words, "It is simply marvellous how a vessel can pass through such an accumulation in the dark and escape with only a few bumps." He is equally laconic about one of the greatest trials of the navigator in the seas near the magnetic pole. "At midnight the engines were reduced to half-speed as the ice looked closer to southward (and) the compass had become sluggish again."

A typical account of the weather describes the day when Mawson was successfully picked up. "At 8 a.m. the land became invisible owing to the driving spray and drift. At 10 a.m. the wind averaged about 70 miles an hour, with squalls of terrific violence. At 11 it reached the strength of a hurricane, the sea was cut off almost flat by the force of the wind, the glass has fallen three-tenths of an inch." (It will interest Melbourne readers to know that the highest single gust recorded at Melbourne only reached a velocity of 69 miles). We are not surprised that the Aurora broke two anchors and lost three others in these tempestuous seas.

Later chapters describe the variations in the Ice Pack off Antarctica. It is most interesting to know that 1914 was marked by an unusually wide and unbroken belt of pack ice. Is it not possible that this greatly affected the temperature of our Australian waters and was a vital factor in determining the great drought of that year? In this section the book is specially well illustrated with sketch maps.

Davis is generous in his praise of the French and American expeditions of 1840. It is interesting to see how poor a showing was made by the Germans in 1902 in these waters, as compared with the explorations of Mawson and Davis in the same region.

In the last chapter he sounds a note of warning. "To the explorer who has not the money to provide good equipment of every kind, my advice is—keep out of the Antarctic!" A final note, and one of the but too few personal touches, is of special interest. "The Australians proved born explorers, though without polar experience. They were always keen, always resourceful, and never cast down by difficulties."

This most valuable oceanographic work is published by Melrose, and is illustrated by nearly 150 photographs and maps. Every Australian who desires to know not only of perilous work to the South'ard, but also something of the chief factors contributing to his own environment, should certainly read this volume.

GRIFFITH TAYLOR.

Ingram Content Group UK Ltd.
Milton Keynes UK
UKHW050844180423
420272UK00010B/19